Biochemical Monitoring of the Fetus

Molly S. Chatterjee
Editor

Biochemical Monitoring of the Fetus

With 29 Figures

Springer Science+Business Media, LLC

Molly S. Chatterjee, M.D.
Dept. of Obstetrics & Gynecology
University of New Mexico
Albuquerque, NM 87131
USA

Library of Congress Cataloging-in-Publication Data
Biochemical monitoring of the fetus / Molly S. Chatterjee, editor.
   p.   cm.
   Includes bibliographical references
   ISBN 978-0-387-97892-5   ISBN 978-1-4757-2259-8 (eBook)
   DOI 10.1007/978-1-4757-2259-8
   1. Fetal blood—Analysis—Congresses.   2. Transcutaneous blood gas
monitoring—Congresses.   3. Laser spectroscopy—Congresses.
   4. Fetal monitoring—Congresses.   I. Chatterjee, Molly S., 1943–
   [DNLM: 1. Fetal Monitoring—methods—congresses.   2. Fetal Blood—
congresses.   3. Cardiotocography—congresses.   WQ 209 B615 1993]
   RG628.3.B55B56   1993
   618.3'207561—dc20
   DNLM/DLC   93-1477

Printed on acid-free paper.

© 1993 Springer Science+Business Media New York
Originally published by Springer-Verlag New York, Inc. in 1993

Production managed by Natalie Johnson; manufacturing supervised by Vincent Scelta.
Camera-ready copy provided by the editor.

9 8 7 6 5 4 3 2 1

ISBN 978-0-387-97892-5

Contents

Preface

Biochemical monitoring of the fetus has been in the back of every perinatologist's mind. Technological advancements have been made in the last ten years but not to the expected level. A continued interest in the subject can only be maintained by symposiums of this nature where perinatologists from different countries can share their experience. Laserspectroscopy of the fetus is a valuable addition to this volume. The future of biochemical monitoring of the intrapartum fetus depends on the continued collection of scientific data and further technological advances.

This successful symposium was held in October, 1990, in Albuquerque, New Mexico, USA. I would like to thank Hewlett Packard for their generous support without which this publication would not have been possible. My sincere thanks goes to my secretary, Nancy Whalen, who has done a tremendous job with the word processing, organization, and layout of the chapters.

Molly S. Chatterjee, M.D.
Associate Professor
University of New Mexico
Department of Obstetrics & Gynecology

CLINICAL IMPORTANCE OF BIOCHEMICAL MONITORING OF THE FETUS DURING LABOR WITH DEMONSTRATION OF TYPICAL CASES

E. SALING, J. BARTNICKI

Institute of Perinatal Medicine,
Free University of Berlin, Berlin, Germany

The biochemical monitoring of the fetus during labor is historically the oldest part of prenatal medicine. The very first direct approach to the human fetus took place on June 21, 1960 when the first blood samples were taken from the fetal scalp in our labor room (3). In the meantime, this method has gone through a typical evolution of ups and downs.

It started with enthusiasm because it was the first opportunity to examine fetal blood samples directly using more or less all clinically interesting analytic laboratory methods. Then when cardiotocography became available in 1968 for clinical routine, many clinicians converted to thinking that this was the philosopher's stone and they renounced on the additional use of fetal blood analysis and relied on cardiotocography alone. How essential this fault was has been demonstrated during the long lasting public discussions about all the adverse effects of one-sided intensive apparative supervision of the fetus during labor.

Now, after several years, most obstetricians have returned to their objective senses and they know that the best compromise is combined supervision during labor, namely the biophysical and biochemical, in other words the use of cardiotocography combined - if necessary - with fetal blood analysis. The latter is now as before the most proved and most widespread biochemical method, which does not exclude that in the future other non-traumatic and maybe not invasive methods will also be applied in the daily routine. <u>What are the specific benefits of both types of the methods used up to now?</u>

1. <u>Cardiotocography</u> has the advantage that it allows a continuous electronic monitoring of the fetal heart rate. If the cardiotocogram is normal an undisturbed condition of the fetus is reliably confirmed. This is an important fact. If on the other hand the cardiotocogram is abnormal, this is - in cases of real intrauterine

complications - a very early sign of threatened hypoxia. But often there is no hypoxia and therefore no dangerous condition.

And so

2.   A biochemical method - for widespread practical clinical use now as before fetal blood analysis - is necessary, which should only be employed as a complementary measure in cases with suspicious or pathological cardiotocogram.

An additional biochemical method enables us:

a.   To clarify whether or not imminent fetal hypoxia and/or acidosis are really present; and if there is no evidence of hypoxia, unnecessary interventions can be avoided. If, on the contrary, early stages of real hypoxia are present, the best suitable time for termination of labor can be ascertained by using biochemical methods.

b.   To recognize whether in cases with imminent hypoxia conservative therapeutic measures are successful, such as use of tocolytics for inhibition of uterine contractions. As it has been shown in an evaluation from 1982 by U. Zitzelsberger (5) in our department in about 75% of all cases with threatened acidosis a conservative treatment with tocolysis was successful in preventing a further fall of fetal pH values. If intrauterine hypoxia progresses in spite of conservative treatment, termination of labor by operation is indicated.

In this way by using both cardiotocography and if necessary fetal blood analysis, it is possible to avoid unnecessary interventions and thus to achieve minimum of operative deliveries. All this can be done without reduction of fetal safety.

The task of modern intensive supervision during labor is certainly not only to avoid occasional severe late brain damage. Too many examiners, unfortunately rely on this one-sided aspect. The main aim of intensive supervision today is to reduce early morbidity to a minimum, particulary in the period shortly after delivery. Severe stage of hypoxia, acidosis and clinical depression must always be taken seriously; they are known to be associated with a number of risks and disadvantages.

Some of the important ones are:
- intracranial hemorrhages occur more frequently
- premature babies have respiratory distress syndrome more often
- the intracellular metabolism is inhibited
- the hematocrit is increased and consequently the blood flow properties are disturbed
- the circulation of numerous organs is reduced
- the breathing center can be depressed, and there can be a delay in the onset of respiration and lung function can be reduced
- fall of cardiac output and a drop in blood pressure
- disturbances of electrolyte balance between the intracellular and extracellular fluid occur
- there is a reduction of the O_2 binding capacity
- renal function is impaired
- transfer of local anesthetic substances from the mother to the fetus is increased.

Another benefit is that the consequent use of biochemical methods can prevent unnecessary risks to the mother. Risks of unnecessary operative interventions are too often underestimated or even ignored. In many places one has got so accustomed to the specific risks of operative measures, for example, cesarean sections are so frequent in some places - that laymen and even many obstetricians consider these risks as being practically non-existent.

From the literature it is known that mortality after cesareans can be 5-10 times higher than after vaginal deliveries. Furthermore, it is known - which is in my opinion much more important - that morbidity after cesareans is also much higher than after vaginal deliveries. Frequency of morbidity up to 30% or more is published (1). So in any case, it is not justified to ignore such risks. As we calculated, the use of biochemical methods in addition to cardiotocography can reduce the absolute rate of cesareans by 1%. If we pragmatically calculate how many unnecessary cesareans these are in a country as a whole, the results look quite serious.

In our country (West Germany), with about 600,000 deliveries per year, the number of unnecessary cesareans would be as much as 6000 avoidable operations, with all the accompanying morbidity and mortality; and in the United States the number concerned should be at least 20,000 avoidable interventions. I think such facts should not be ignored.

From all that has been said, the logical conclusion is, that at all places where a biochemical method is not used in addition to cardiotocography, obstetrics is not performed in a progressive way. After monodiagnostic cardiotocographic indication, operative intervention would have been necessary in 73% of such high risk cases concerned, in order not to miss the 14% of fetuses who really were at high risk due to a fall in pH values. So here we have an over-diagnosis through cardiotocography of around 60%. We found such results in a previous evaluation in our unit together with K. Goeschen and T. Gruner (2). If the first stage and the second stage of labor are subdivided, the following picture emerges: in the first stage an operative delivery - mostly a cesarean due to suspicious or pathological cardiotocograms - would have indicated in 56% of the cases. After performing fetal blood analysis, however, the cesarean rate was only 10% - this means a saving of 46%.

Also in the second stage decisive advantages were found when combined monitoring was used. In a high risk group a considerably abnormal CTG-score after Hammacher was recorded in 95% of the cases. Instead of having to operate on all these 95% high risk cases, through the results achieved by fetal blood analysis it was only really necessary to make an operative intervention in 54% of them. The remaining 41% infants could be delivered spontaneously because of not too critical pH values. According to our experience fetal blood analysis is indicated in the early second stage, when the cervix is completely dilated but the presenting part is in the mid-pelvic plane or still higher.

Operative interventions in the early second stage - that is mainly from the mid-pelvic plane, can by no means be regarded as harmless for the fetus. Often they require difficult manipulation and should therefore only be performed when there is a strong and real clinical indication. Operative interventions should not be performed when the heart rate patterns are suspicious or pathological but not due to hypoxia. In a prospective study in our department performed together with M. Brand, we were able to show that intracranial hemorrhage - mostly of a slight degree - occurs twice as frequently in mature infants delivered operatively in the early second stage (10.5%) than in the late second stage (4.5%) and almost four times more frequently than in infants delivered spontaneously where the incidence was 2.6%.

False interpretation of the reliability of intrapartum monitoring
Let us see what happens when fetal hypoxia starts, from another viewpoint. A cardiotocogram is suspicious and fetal blood analyses are performed. The pH-values are still within the normal range but are sinking and cesarean is performed in good time. The pH-values measured immediately after delivery in the umbilical artery blood are still just within the normal range and the baby is clinically vigorous.

In various publications such cases are erroneously classified as "false positive"; this is because the newborns are not born in pathological condition. The examiners concerned have apparrently forgotten an important fact, namely to ask themselves what is the task of modern method of supervision during labor at all? The task cannot only then be considered as having been achieved when - after suspicious findings immediately after delivery - pathological conditions are always present in the form of acidosis and/or a clinical depression of the baby.

From the pathophysiological aspect it must be clear that a considerable number of infants still do not show signs of acidosis and are clinically vigorous. From clinical and scientific aspects it is not acceptable to classify the findings in those infants as "false positive".

We have performed in our unit a study concerning these questions together with Sabine Brandt and P. van den Berg. Out of 110 fetuses who had abnormal cardiotocograms and simultaneously reduced scalp pH-values, 94% had, immediately after delivery an equal or lower pH-value in umbilical artery samples. Only 6% had higher pH-values. This means that fetal blood analyses have been misleading in only about 6% and the measured values would have been labelled as "false positive". So even with biochemical monitoring single unnecessary interventions are possible but rare.

Concerning the question whether it is recommendable to intervene operatively when fetal scalp pH-values are falling below 7.25 or when they are already acidotic at the first FBA - in both cases this is the rule in our clinical routine - we also have some important results (Fig. 1).

It is particularly interesting that as many as 62 out of the 110 infants, thus 56% were born in an acidotic state and 27, thus 24% in a clinical state of depression; this shows that the clinician has no time to lose waiting until there are much lower fetal pH values, because otherwise he would have to expect more cases with severe acidosis and severe depression. We consider an infant to be seriously endangered by hypoxia when it is born in an advanced or severe state of acidosis (pH < 7.1) combined with an advanced or serious clinical state of depression (Apgar 4 and less). The number of such infants was 4 out of 110, thus 3.6% of this high-risk group which I just mentioned. In the total material of our department, the number of such severely endangered newborns was at the same time 20 out of 5854 born infants, thus only 0.3%. So in our study the incidence of acidotic newborns was 10 times and of depressed newborns 8 times higher than in the total number of infants born.

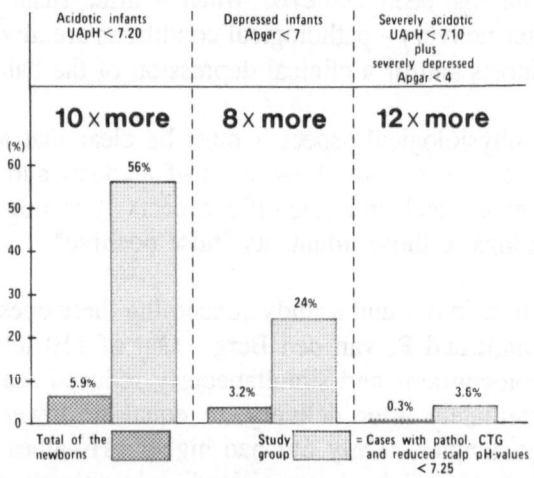

Figure 1: This comparison shows that it is not advisable to wait any longer than we do, namely when the pH values are apparently falling or are apparently reduced. Such facts are in so far important as several clinicians sometimes have their doubts whether the borderline between normality and pathology as we have drawn at 7.2 and more for normal and 7.19 and less for pathological pH-values is justified. Our results show that such a borderline corresponds with clinical aspects very well.

Demonstration of some typical clinical cases concerning the use of biochemical methods during labor

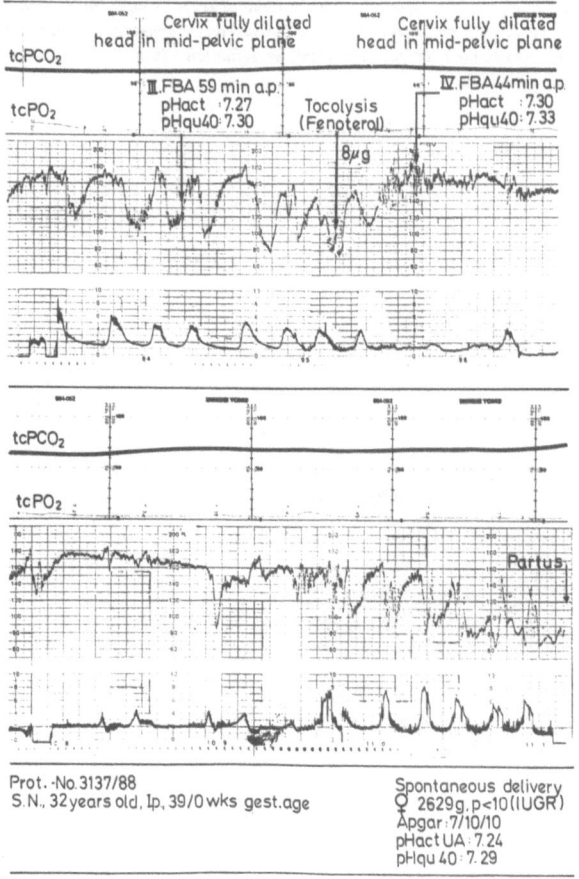

Case 1, Figure 2 (Protocoll Nr. 3137/88)

Figure 2: A 32 year-old nullipara went into labor with 39 completed weeks of gestational age. One hour before delivery the cardiotocogram became pathological. Pronounced variable decelerations appeared when the cervix was fully dilated and the head was in the mid-pelvic plane. Continuously measured tcPCO₂ values had been within the normal range, confirmed by fetal blood analysis. Actual pH was 7.27 and pH after equilibration with 40 mm Hg, PCO₂ - the so called pHqu40, which represents the metabolic acidity - was 7.30. The recorded trancutaneous PO₂ was performed here not to get any information about the oxygen situation in the fetus but only to confirm with the so low oxygen tension that the electrode for transcutaneous measurement is correctly fixed on the skin of the fetus.

Because of the slightly reduced but still not pathological actual fetal pH value, tocolysis with 8 ug Fenoterol has been performed intravenously. The labor activity decreases and the next blood sampling 44 minutes ante partum showed normal pH values from the fetal scalp. Forty minutes later a vigorous slightly growth retarded *baby below the 10th percentile was born spontaneously with still normal pH values in the umbilical artery.*

Our case demonstrates that without the additional biochemical supervision - in this case with tcPCO$_2$ recording and with fetal blood analysis - an operative termination of labor would have been performed by many obstetricians. In any case, such an operation, either a cesarean or a viginal operative extraction of the baby has - as already mentioned in the previous part of this presentation - its negative side-effects and risks for the mother, for the fetus or for both.

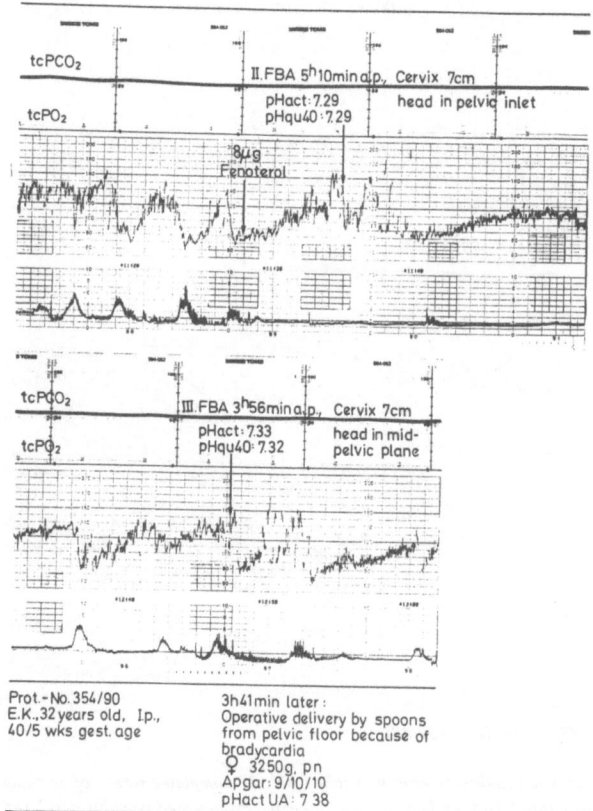

Case 2, Figure 3 (Protocoll Nr. 354/90)

Figure 3: 32 year-old nullipara, 40 weeks plus 5 days gestational age started spontaneously with labor activity. Five hours and 10 minutes ante partum, severe variable decelerations with reduced varability during contractions appeared accompanied by bradycardia. The cervix was dilated 7 cms and the head was in the pelvic inlet. The tcPCO$_2$ values and fetal pH were in the normal range. After tocolysis with 8 ug Fenoterol, applied intravenously, the labor activity was reduced and the cardiotocogram became better. Another episode with variable decelerations happened later namely 3 hours and 56 minutes before delivery. Biochemically the fetus with an actual pH of 7.33 was not impaired. So after the next period of about 4 hours, the baby could

9

be delivered by so called spoons (modified forceps), from the pelvic floor because of the terminal bradycardia. At this late stage of labor, we do not perform more fetal blood analysis because the fetus can be delivered immediately without remarkable additional risk.

Also in this case the additional biochemical information gave the obstetrician the assurance that, in spite of a pathological cardiotocogram, the fetus was not in danger through hypoxia and the labor could be continued in a normal way until a vaginal delivery from the pelvic floor became possible.

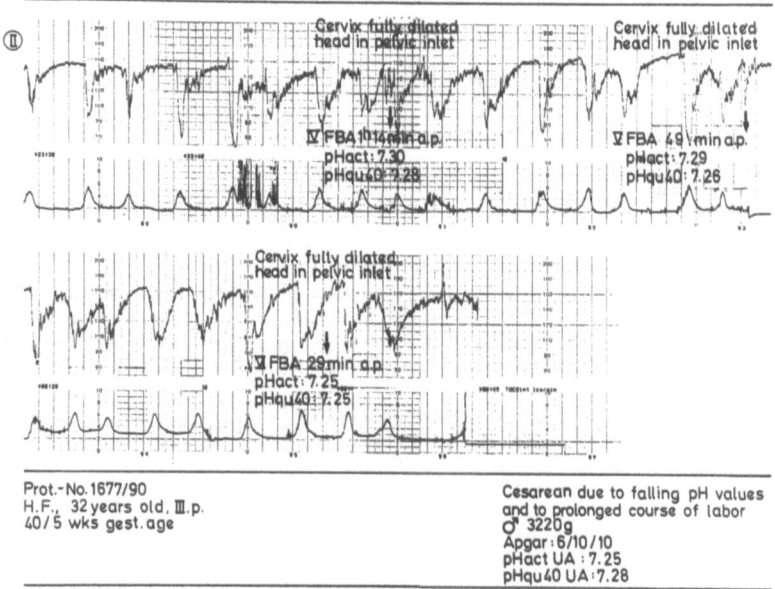

Prot.-No. 1677/90
H.F., 32 years old, III.p.
40 / 5 wks gest. age

Cesarean due to falling pH values
and to prolonged course of labor
♂ 3220g
Apgar : 6/10/10
pHact UA : 7.25
pHqu40 UA : 7.28

Case 3, Figure 4 (Protocoll Nr. 1677/90)

During labor of a 32 year-old para II with 40 weeks plus 5 days of gestational age, one and a half hours before delivery the cardiotocogram became again pathological with pronounced variable decelerations. Fetal blood analysis performed 1 hour and 14 minutes before delivery initially showed normal pH values. As the cardiotocogram remained pathological after 25 minutes and again after 20 minutes fetal blood analysis was repeated. pH values had shown a slight falling tendency. As the labor was prolonged with no progress during the last hour, a cesarean was performed. A slightly depressed infant (probably due to general anesthesia) (Apgar 6) with still normal pH values in the umbilical artery blood was delivered.

In this case, the biochemical supervision could not achieve any clinical benefit because of the necessity of performing a cesarean caused by another reason.

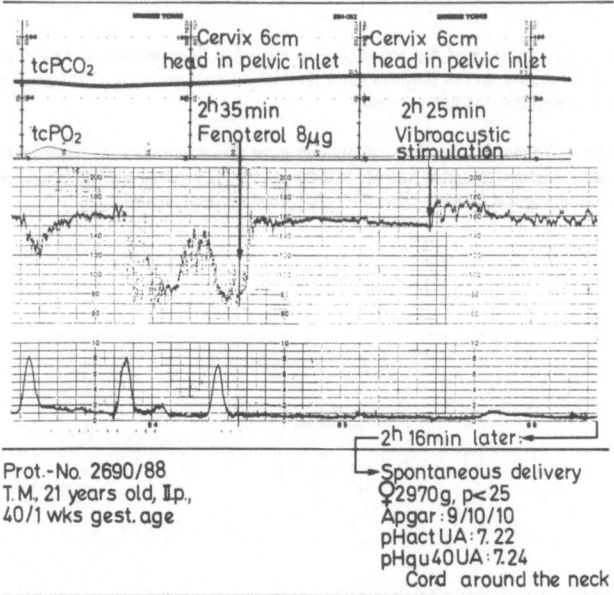

Figure 5: The fetus of a 21 year-old para I was supervised by continuous tcPCO₂ measurements. When two prolonged decelerations appeared, which came close to a bradycardia, 8 ug Fenoterol was applied which stopped the contractions for a while. Ten minutes later, vibroacustic stimulation instead of FBA followed. The fetus reacted promptly. So we could consider that the fetus was in an undisturbed condition.

In this connection it should be mentioned that vibroacustic stimulation as proved particularly by Smith and co-workers in 1986 (4) can reduce the frequency of fetal blood analysis to a considerable extent. After an evaluation together with our co-worker, W. Maeckert,

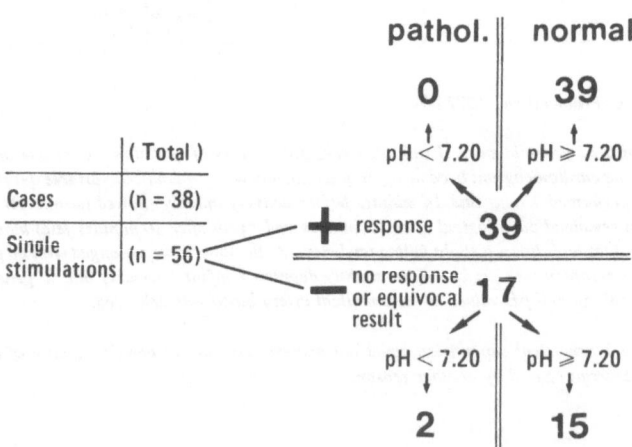

Figure 6: in 38 cases with 56 vibroacustic stimulation tests performed with an artificial electronic larynx we

found that in all 39 single tests with positive response of the fetus - this means an acceleration of 10 or more bpm - there has never been a fetal scalp pH-value of less than 7.20. The conclusion is that vibroacustic stimulation has a high safety to exclude critically reduced pH-values. Out of 17 tests with absence of reactive accelerations or with an equivocal result - for instance difficult differentation to spontaneous short term variability - 15 times the fetal pH was also normal, and only in 2 tests reduced. So we can draw the conclusion that in all cases without fetal heart rate acceleration or in cases with an equivocal result, fetal blood analysis should be performed to prevent unnecessary operative interventions.

We come back to our case. Two hours and 16 minutes later this infant was born spontaneously in a vigorous state with still normal pH values in umbilical artery blood. The cord around the neck was the probable reason for the pathological heart rate patterns.

Results of combined intensive supervision during labor. *Today our special interest concerns the results achieved through modern intensive supervision during labor by cardiotocography and fetal blood analysis. The mortality during labor, assessment of the newborn immediately after delivery and the cesarean rate play the main role for this consideration. Our fetal mortality rate during labor during the last 22 years (Figure 7), since we*

Mortality during labor				
Methods of super-vision	Only auscul-tation	Ausculta-tion and FBA	CTG + FBA	
Year	1955–1960	1961–1967	1968–1977	1978–1989
Number of born infants	n= 10431	n= 18326	n= 20812	n = 34896
Number of fetal deaths	n = 58	n = 58	n = 36	n = 15
%	**0.6**	**0.32**	**0.17**	**0.04**
Signific.	$p < 0.005$ $p < 0.001$ $p < 0.005$			
	⟶ fifteen times lower ⟶			

Figure 7: *have used cardiotocography as a routine measure in nearly all cases combined with fetal blood analysis, averaged 0.09%; during the last 12 years even 0.04%. This is minimal particularly if we compare it with the mortality during labor before introduction of fetal blood analysis when the fetuses have been supervised only by auscultation with the simple stethoscope. This mortality was 0.6% thus 15 timed higher than now. The cesarean rate during the past twelve years in our department has remained between 8 and 10%, which is quite low. The frequency of fetuses and newborns seriously endangered by hypoxia - this means Apgar less than 4 combined with an umbilical artery pH less than 7.1 - has ranged during the last twelve years between 0.1% and 0.3% and were thus also very low.*

General review of intensive supervision during labor

The use of intensive supervision of the fetus during labor should also be regarded form a more superordinated point of view. We know that even uncomplicated birth process is a relatively short period of concentrated occurences, which nowadays take their course within, lets say more or less 12 hours. In no other period of our life are we threatened by such concentrated risks in such a short space of time. Futhermore, we have to take into account that in the industrial countries today a woman on an average lives for 77 years, that is around 28,000 days. If we compare this with the duration of labor - roughly 12 hours as mentioned above (but often much shorter) - we get a ratio between the duration of intensive supervision against the whole span of life of around 1:56,000. This is the same relationship if one would compare the length of a ladybird with the height of the Eifel Tower in Paris. Should it really be considered worth it, to do without intensive supervision for a 56 thousandth part of life and to pay attention exclusively to our emotions and psychic needs and thus naively neglect the safety for such an extremely short period with the most concentrated threat of danger? I think the best solution for the future will be to recommend and to offer all patients both the high safety of modern intensive supervision during labor such as described in this presentation combined with good individual and personal psychological care, which more and more progressive obstetrical departments are successfully achieving.

References

1. Amirikia, H., B. Zarewych, T.N. Evans: Cesearan section: a 15-year review of changing incidence, indications and risks. Am. J. Obstet. Gynec. 140 (1981)81.

2. Goeschen, K., T. Gruner, E. Saling: Stellenwert des Hammacher-Scores und der Fetalblutanalyse bei der subpartualen Uberwachung des Kindes. Z. Geburtshilfe Perinataol. 188 (1984) 12.

3. Saling, E. In Rooth, G. O.D. Saugstad: The Roots of Perinatal Medicine. Thieme, Stuttgart 1985.

4. Smith, C.V., J.P. Phelan, L.D. Platt et al.: Fetal acoustic stimulation testing. II A randomized clinical comparison with the non-stress test. Am. J. Obstet. Gynec. 155 (1986) 131.

5. Zitzelsberger, U.: Tokolyse sub partu. Doctoral Thesis. The Free University of Berlin, 1982.

ANIMAL STUDY AND CLINICAL APPLICATION OF LASERSPECTROSCOPY IN THE FETUS

S. Schmidt, S. Gorissen-Bosselmann, S. Spaniol, U. Wagner,
K. Pringle, N. Helledie, P. Rolfe, D. Krebs

Women's Hospital, University of Bonn, Germany
Department of Biomedical Engineering and Medical Physics,
University of Keele, England
Department of Paediatric Surgery,
University of Wellington, New Zealand

INTRODUCTION

The biochemical monitoring of the fetus has been started about thirty years ago, when Saling introduced the technique of fetal blood analysis into clinical routine (1,2,4,7,13,14,15). An improvement of the fetal surveillance as well as a better understanding of the pathophysiology of the fetus were achieved by intermittent analysis of pH, pCO2, pO2 and lactate concentration (11,15,21,22). Today non-invasive techniques for continuous tracing became available and were evaluated in in-virto studies. The near infrared laser spectroscopy provides information about biochemical parameters by the analysis of rescattered laser-light (3,9,13,18,19,20).

The principle of relectance spectrophotometry was first described by Jobsis about the non-invasive measurement of cerebral and myocardial oxygen sufficiency and circulatory parameters by the means of near-infrared spectroscopy(9).
The introduction of laser-diodes and the development of mathematical procedures as proposed by Peter Rolfe led to configurations of measuring systems that can be used during research routine(13).

One objection against the technique was due to the fact that it is based on the measurement of overlapping signals from two biochemical substances - namely hemoglobine and cytochrome aa3. It was thus the aim of our animal studies to prove the ability of laser spectroscopy in differentiating the changes of the intracellular redox state respectively the oxygen saturation in the blood.

Monitors for laserspectroscopy are available for clinical research routine and have been applied in the field of neonatal intensive care(3). During fetal monitoring a continuous record of a biomedical indicator is preferable. The configuration of the measuring system aims at the non-invasive calculation of both the intracellular redox-state and relative changes of blood volume. For the observation in the fetus a modification of the application system was necessary. We will report on clinical trial during fetal monitoring.

Monitoring system
The Radiometer prototype that was used comprises a personal computer (Hewlett Packard - Vectra) interconnected with a near-infrared data collection unit (NIRDCU). This unit includes four pulsed laser diodes (775,805, 845, 904 nm), and a special constructed microcomputer board for data processing. Additionally a Photomultiplier tube acts as optical reciever for detecting the reemitted laser light. The light of the 4 laser diodes and the reflected light is delivered by two plexiglass fibers connected to glass prisms. Changes of the density in relation to the time are displayed on a colormonitor, while data are stored for later processing(17).

ANIMAL STUDY

The animal study was performed using five fetal lambs (130 days gestation). The preparation of the animals was performed after laparatomy and uterotomy (6,10,14). A canulation of both the Jugular vein and the Carotid artery ws performed for the connection with a veno-arterial extracorporeal bypass (Fig. 1). The extracorporeal circuit consisted of silicone tubing, a reservoir, a mechanical pump and a Scimed - Kolobow membrane lung. The extracorporeal circuit was primed with 100 cc of fluorcarbon-Ringerlactate solution (70%). Room air was provided by a suction at the gas outlet port of the membrane lung. After heparinisation of the lambs (3mg/kg) extracorporeal perfusion was performed with a flow rate of 100 cc/min providing a fluorcarbon - blood exchange in the cerebral vessels(13).

Laser prisms were placed on the fetal scalp and perfusion was started only when signals were stable and fetal blood sample was confirming a normal acid balance (6).

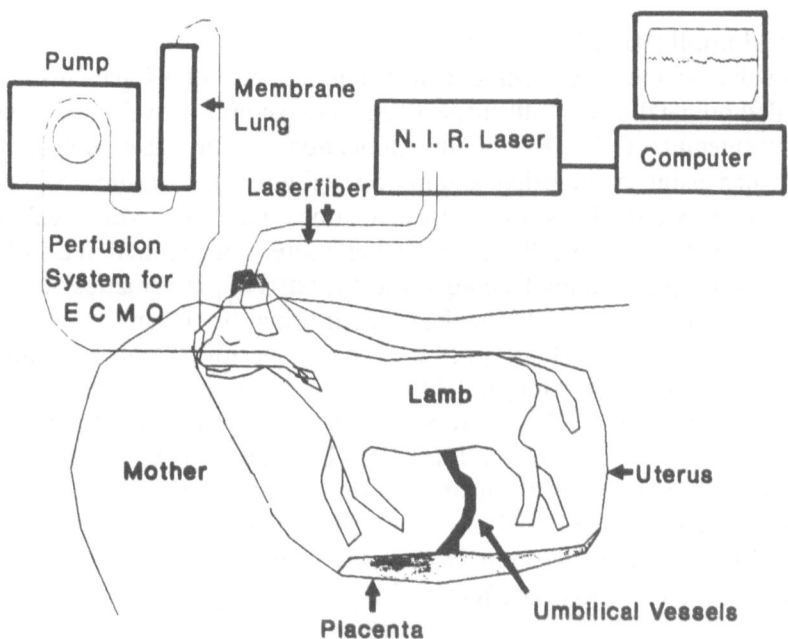

Figure 1: Schematical drawing of the experimental setup for fluorcarbon infusion using an extracorporeal circuit (ECMO) during laser spectroscopy in fetal lambs. Laserfibers are placed on the fetal head during volume exchange via the Carotid artery and Jugilar vene. Umbilical circulation is undisturbed until induced compression.

During the animal study we achieved a characteristic tracing showing stable values with only minor fluctuations of oxidized hemoglobine (HbO2), reduced hemoglobine (HbR), calculated blood volume and cytochrome aa3 (Fig. 2, see Color Plate 1).

With the infusion of fluorcarbon a reproducible change was induced. After a neglectible lagtime of less than two seconds a fall in the HbO2 signal and a rise in HbR. On the other hand the cytochrome signal stayed stable until cord compression was induced (5,10). This result implicates that the mathematical compensation for the change of the hemoglobine signal is sufficient. This allows a reliable detection of relative cytochrome aa3 changes.

Clinical application in the fetus
We evaluated the new technique in a clinical trial of 94 patients. We included prenatal cases with suspect heartrate patterns as well as sub partu measuremenmts (17, 18). The application of the laser sensor was performed using a modified amnioscope (figure 3). Cardiotocography with a Hewlett Packard autocorrelation monitor was evaluated semiquantitatively using the Hammacher score. Sequentiell fetal blood sampling using the Saling technique was the basis of mathematical analysis of the accuracy of the biomedical monitoring by means of a linear correlation function. Blood gas analysis was performed on a automatic blood gas analyser (Corning). By using the transvaginal route we were not able to achieve a reliable tracing of the biophysical and biochemical analysis (Figure 4, see Color Plate 2). In cases with given suspect of pathologic CTG tracings, the near infrared spectromter signal stayed stable in the majority of cases (97%). The correlation between PO_2 values of fetal blood analysis and the HbO_2 was statistically significant. We calculated a correlation coefficient of $R = 0.96$ (figure 5).

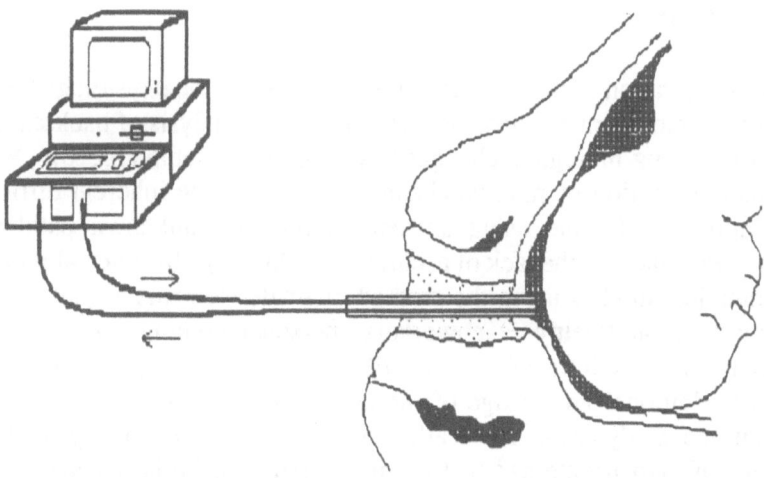

Figure 3: Fetal application of laserspectroscopy: laserfibres are placed on the fetal head using a modified amnioscope.

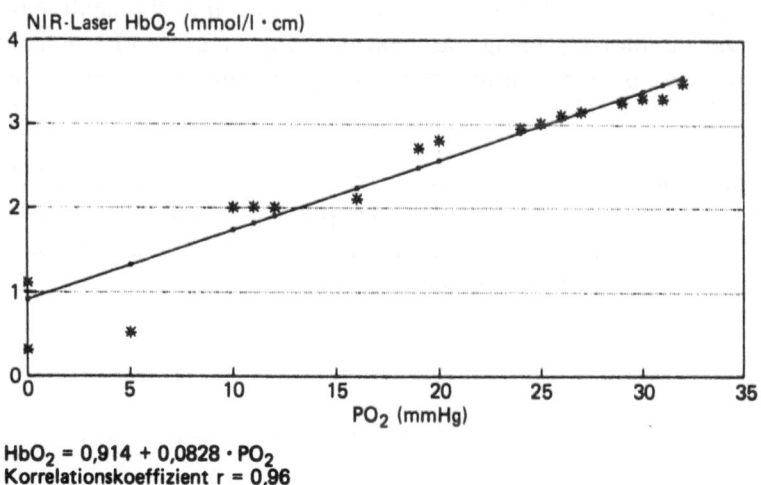

$HbO_2 = 0{,}914 + 0{,}0828 \cdot PO_2$
Korrelationskoeffizient r = 0,96
n = 16

Figure 5: Correlation of pO_2 values from fetal blood analysis with HbO_2 of the laserspectrometry. A significant correlation with a linear coefficient of R = 0.96 was calculated.

DISCUSSION:

After inauguration of near-infrared-spectroscopy by the physiologist Jobsis and the development of adequate laser diodes the analysis of backscattered light for tracing biological changes has become available (13,20). While a reliable detection of relative changes is possible the inherent difficulty of a aptical technique using scattered light with unknown pathlength (Labert-Beer-law) is the lack of accuracy in achieving absolute values (13). mathematical models using the method of finate elements are in the state of investigation (Delpy et coworkers, personal communication). The ability to trace the intracellular state of oxygenationis of special interest to achieve additional knowledge of the pathophysiology of cerebral palsea. Our animal study using fluorcarbon implicates the possibility to detect changes of cytochrome aa3 that are independent of initial changes of the blood situation (5,13). The observation of a considering lack time of the reduced intracellular redox state after cord compression fits with the observation using electroencephalography (10,18). While cardotocography is widely accepted as the basic method of fetal monitoring it has the obvious shortcoming by giving false positive indications of fetal distress leading to unnecessary deliveries (16). Here the clinical potention of non-invasive monitoring using laserspectroscopy lies in the potential to discriminate between time respectively contration related changes (7,17,18).

In conclusion, we might say that in the furture near-infrared-laserspectroscopy using relative unexpensive mobile monitors might become a valuable clinical tool in the detection of fetal complications improving the safety for mother and child.

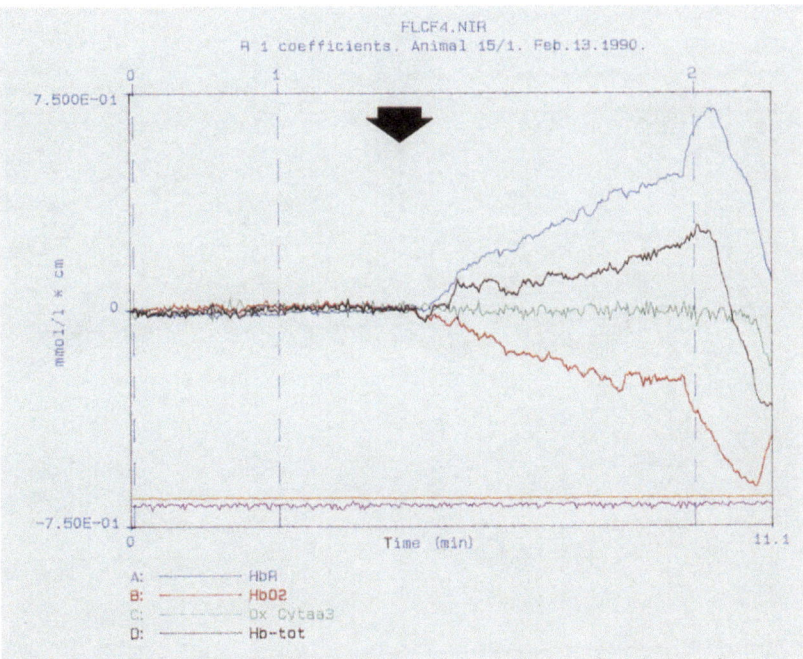

Color Plate 1, *Figure 2: Laserspectroscopic tracing during fluorcarbon infusion in fetal lamb. With a short in vivo reaction time blood desaturation is indicated both by a rising HbO₂ signal. Cerebral oxygenation is undisturbed as indicated by a stable cytochrome aa₃. The terminal decline is induced by a clitical disturbance of umbilical circulation after cord compression. (1) Blood gas analysis confirming normal status of acid blood balance. (↑) Start of fluorcarbon infusion. (2) Umbilical cord compression.*

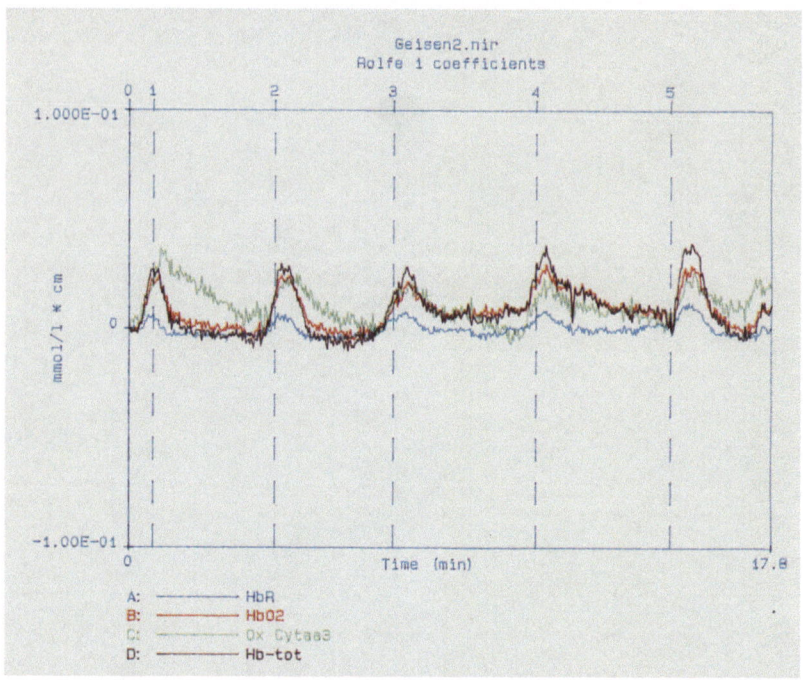

Color Plate 2, *Figure 4: Laserspectroscopic trasing during fetal monitoring: Oxygenated hemo-globine (HbO$_2$), reduced hemoglobine (HbR), calculated blood volume and cytochome aa$_3$ as calculated from the relative absorbances at 775, 805, 854, 904 nm by means of the Rolfe coefficients. During contractions (1-5) a significant rise of near infrared parameters is induced. The original value is reached in the interval to the following contraction.*

REFERENCES

1. Brazy, J.E., Lewis, D.V., Mitnick, M.H., Jobsis van der Vliet, F.F. Non-Invasive Monitoring of Cerebral Oxygenation in Preterm Infants: Preliminary Observations Pediatrics (1985), 75, 2:217-225

2. Cady, E.B., Dawson, M.J., Hope, P.L. et al: Non-invasive investigation of cerebral metabolism in newborn infants by phosphorus nuclear magnetic resonance spectroscopy Lancet (14.5.1983) 1059-1062

3. Chance, B. and Williams, G.R.: Respiratory enzymes in oxidative phosphorylation J. Biolog. Chemi. (1955), 217:409

4. Dunn, L.K., Redstone, D., Roe, H.L., Steer, P.J., Beard, R.W. The relationship between tissue and arterial pH in hypercardiac rabbits Arch Gynecol (1978) 226:31

5. Goeschen, K., Gruner, T., Saling, E. Stellenwert des Hammacher-Scores und der Fetalblutanalyse bei der subpartualen Uberwachung des Kindes Z. Geburtsh. u. Perinat. 188 (1984) 12

6. Jensen, A., Kunzel, W. The difference between fetal transcutaneous PO2 during labor Gynecol Obstet Invest (1980) 11:249

7. Jobsis, F.F. Non-invasive, infrared monitoring of cerebral and myocardial oxygen sufficiency and circulatory parameters Science (1977), 198:1264

8. Jobsis, F.F., Keizer, J.H., LaManna, J.C. et al: Reflectance spectrophotometry of cytochrome aa# in vitro. J. Appl. Physiol. (1977), 43:858-872

9. Jobsis, F.F.: Oxidative metabolic effects of cerebral hypoxia Advances in Neurology (1979), 26:299

10. Nickelsen, C., Weber, T. Continuous transcutaneous carbon dioxide monitoring during human labor J. Perinat. Med. (1988)16: 107-111

11. Rea, P.A., J. Crowe, Y. Wickramasinghe and P. Rolfe Non-invasive optical methods for the study of cerebral metabolism in the human newborn: a technique for the future? Journal of Medical Engineering & Technology 9 (1985) 160

12. Romer, V.M., Kieback, D.G., Buhler, K. Zur Frage der fetalen Uberweachung sub partu in der Bundesrepublik Deutschland Geburtshilfe Frauenheilkde (1985) 45: 147

13. Saling, E. Fetal scalp blood analysis. J. Perinat. Med. 9 (1981) 165

14. Saling, E. Pathophysiology, clinical relevance of continuous measurements of pH and/or CO_2 in the fetus J. Perinat. Med. (1984) 12: 234

15. Saling, E. Introduction and clinical aspects of biochemical monitoring of the fetus J. Perinat. Med. (1988) 16:23

16. Schmidt, S. Methodology and clinical value of transcutaneous blood gas measurements in the fetus. J. Perinat. Med. (1988)16:95

17. Schmidt, S., Eilers, H., Lenz, A., Helledie, N., Krebs, D. Laserspectroscopy in the fetus J. Perinat. Med. 16 (1989)

18. Schmidt, S. Gorissen, S., Eilers H., Decleer W., Krebs, D. Laserspektroskopie - Ein neuses Verfahren zur Uberwachung des Feten Geburtsh. u. Frauenheilk. 50 (1990) 344-348

19. Schmidt, S., Decleer, W., Gorissen-Bosselmann, S., Pringle, K., Helledie, N., Rolfe, P., Krebs, D. Die Belastung del fetalen Gehrins bei Nabelschnurkompression - eine tierexperimentelle Studie mittels Laserspecktroskopie Z. Geburtsh. u. Perinat. 194 (1990) 219-223

20. Schmidt S., Decleer, W., Gorissen-Bosselmann, S., Eilers, H., Pringle, K., Helledie, N., Rolfe, P., Krebs, D. Laserspektroskopische Erfassung der induzierten Hyperoxie - eine tierexperimentelle Studie beim Lamm Biomedizinische Technik, Band 35, Heft 9/1990

21. Schneider, H., Huch, R., Schachinger, H. Correlation between scalp tcPCO2 and microblood samples In: Huch, A., Huch,R., Lucey, J.F. (Eds) Original article series-birth defects The national foundation march of dimes, Liss. New York (1979) 15: 235

22. Sturbois, G., Uzan, S., Rotten, D., Breart, G., Sureau, C. Continuous subcutaneous pH measurement in human fetuses - correlation with scalp and umbilical blood pH Am. J. Obstet. Gynecol (1977) 128:901

COMPUTERIZED FHR ANALYSIS
AND BIOCHEMICAL CHANGES

G.S. Dawes and C.W.G. Redman

Nuffield Department of Obstetrics and Gynaecology, Oxford

ANTENATAL

Over the past 12 years a computerized fetal heart rate analysis system, for antenatal use, has been developed, improved, and recently made commerically available by Oxford Sonicaid. It should replace the non-stress and contraction stress tests because it is more reliable. It has not yet been tested systematically against the biophysical profile.

The system has been in routine clinical use in Oxford since 1982, where we have accumulated more than 17,000 records. We are grateful to colleagues in England who made available a further 6,500 records from a randomized trial, and to Prof. Mandruzatto who sent us another 7,000 records from Trieste, Italy. The system is now the subject of a multi-centre trial by the European Community Concerted Action Project, "New Methods for Perinatal Surveillance." One of the benefits from computerized analysis is that it is objective, while visual analysis of FHR patterns suffers from large inter- and intra-observer variation (1-7). Another advantage has been that thousands of records can be re-analyzed rapidly, to test the effect of program changes in improving discrimination between a normal fetal outcome and otherwise. One of the lessons learned has been that many thousands of records must be scrutinized to ensure that the baseline fetal heart rate (FHR) can be fitted and drawn satisfactorily.

The principle of operation is straight forward. The computer interrogates a FHR monitor (Hewlett Packard 8040 or Oxford Sonicaid FM7) at short intervals (100 msec) in order to record pulse intervals. The system is interactive, warning the midwife if the ultrasound transducer or tocodynameter needs adjustment as judged by excessive signal loss. This led to a 50% reduction in signal loss in randomized trial by Dr. Wheeler in Southhampton. Secondly, the time spent recording is reduced, on average to 15 minutes, and is better allocated. Thus advice is given to stop recording after only 10 minutes if an episode of high FHR variation is detected. But in the absence of such reassuring information, or when for example a large deceleration is detected, recording is allowed to continue up to 1 hour. Thus record length is adjusted to requirements (8).

In a preliminary study (9) we found that the traditional criterion of fetal normality in the non-stress test (i.e. the presence of 3 or more accelerations of amplitude at least 15 bpm and duration 15 secs or more) was unsatisfactory because there were too many false positives. In many fetuses of more than 34 weeks gestation, with normal outcome, this criterion was not satisfied within one hour; the proportion was even greater at earlier gestational ages (17% of records at 28-33 weeks). The criterion of detecting an episode of high FHR variation proved more satisfactory, and has proved reliable even in the absence of accelerations (of amplitude, 15 or 10 bpm).

The program is also designed automatically to detect, and to eliminate from subsequent calculations, brief abrupt decelerations and accelerations, considered spurious because the fall and rise are so rapid. Such errors are common, have usually been attributed to detection of the maternal rather than the fetal pulse, but they are more likely to be due to limitations in the fetal monitors, probably from failures in the application of the autocorrelation process (10). There are such unreliable accelerations as well as decelerations: the incidence of unreliable decelerations varies with the manufacturer, being present in an average of 6-8% of records. It is common sense to eliminate such unreliable features.

An additional feature of the system is the fetal movements, detected by the mother (or midwife) are routinely recorded by pressing a hand-held button.

Using this system for the past 8 years we have observed that:

1.   Flattening of the FHR pattern with intrauterine growth retardation is due to a decrease in variation in episodes of high FHR variation (corresponding to active sleep postnatally), rather than to prolongation of episodes of low variation (quiet sleep, 11).

2.   Moderate, but abnormal, reduction of long term FHR variation in growth retardation is associated with umbilical arterial hypoxaemia without metabolic acidemia (12,13) The hypoxaemia is probably chronic since it is accompanied by a 4-fold rise in liquor erythropoietin. And there is hypoglycaemia and a rise in plasma alanine concentration (as in fetal lambs whose placental mass has been reduced experimentally). Yet there is no significant increase in umbilical arterial plasma hypoxanthine or endorphin concentrations, as in acute asphyxial deliveries.

3.   Further fetal deterioration, usually accompanied by a reduction in fetal movements, is best detected by a decrease in short term FHR variation. The reason for this preference (in contrast to measurement of long term variation) is that occasionally a sinusoidal rhythm may appear terminally. The amplitude (up to 10 bpm or more) and frequency (1 in 2-5 mins) is such that it can contribute to a relatively high measure of long term variation, which would give misleading reassurance. Although such sinusoidal patterns are rare (about 1 in 1,500 records), failure to recognize their sinister significance could lead to an unanticipated intrauterine death.

The normal level of short term FHR variation is about 9 msec. A warning is given when it is reduced to 3 msec, and an additional warning of impending metabolic acidaemia and/or intrauterine death when it falls to 2.5 msec or less. In 9 fetuses, which were not delivered by caesarean section after this latter warning (short term FHR variation < 2.5 msec) because of prematurity (25-28 weeks) and small size (<500 g), intra-uterine death ensued, usually within 18 hours and always within 48 hours.

4.   The presence of repeated decelerations, either large or small and associated with uterine contractions, proved to be a less satisfactory indication of fetal deterioration. Similarly tachycardia did not correlate with metabolic acidaemia on delivery, or intra-uterine death. There are some fetuses, in our series and in the literature, with a very flat heart rate and no decelerations. This occurs in some instances of maternal drug administration, or

(rarely) with fetal central nervous anomalies or cerebral infarction, or in ventricular tachycardia. Hence, the absence of decelerations is a diagnostic factor of importance. Paradoxically the presence of decelerations demonstrates that the lower brainstem is functioning normally and that the heart can respond appropriately.

5.   In fetuses with a somewhat reduced FHR variation, at the bottom of the normal range, there is a rapid rise postnatally (13). But in fetuses which are growth retarded the rise in FHR variation is much less rapid in spite of the large increase in PaO2 after birth. Therefore we suspect that the decrease in FHR variation is not due to a direct effect of oxygen lack. Indeed animal experiments, in both sheep and monkeys, have shown that acute hypoxia causes an immediate increase in FHR variation, attributd to catecholamine release. The decrease in FHR variation in chronic hypoxaemia in anomalous. It could be due to failure of catecholamine release, to down regulation of the receptors or/and to central adaptation to chronic hypoxia as recently demonstrated in fetal lambs (for reference see 15).

Finally, it should be emphasized that the decrease in short term FHR variation, which has proved in Oxford a useful clinical tool, is an empirical observation, though consistent with much previous literature. What our computerized system achieves is measurement, reliable discrimination from natural changes episodically with sleep/behavioral state from 28 weeks gestational age, and warnings to staff (midwives or doctors) who may not yet have had experience. For the incidence of truly abnormal FHR records is low, < 1% in low risk clinics. It needs prospective trial on a large scale, and such a trial is now under way in Europe.

OBSERVATIONS IN LABOR

We have also made some preliminary observations in labor over the past 5 years, using either the antenatal program to screen a population in early labor or a program modified to run continuously, also on-line. The screening study demonstrated a low, but not negligible, incidence (1 in 588) of fetuses already compromised as shown by large decelerations and an otherwise very flat trace (16). This is consistent with a previous report (17) and from further studies we can calculate, tentatively, an incidence of 2-3 in 3,000 deliveries near term (37-42 weeks gestation).

Excluding the fetuses already compromised, further studies throughout labor have shown that episodes of low or high FHR variation (characteristic of sleep/behavioral states) persist in early labor. In some instances they are evident throughout labor. The episodes of low FHR variation are often of very low amplitude and may persist for as long as an hour, yet without metabolic acidaemia on delivery. The significance of this confounding variable has not hitherto been systematically considered. Measurement of FHR variation shows that it increases during labor, on average by 40%, consistent with the increased fetal plasma catecholamine concentration at birth. Since vaginal delivery is associated with a lower PaO2 than on cordocentesis antenatally, this is evidence that acute hypoxia in the human fetus is associated with an increase in FHR variation. Our preliminary results also suggest that baseline FHR is not correlated with metabolic acidaemia, when analyzed objectively by computer.

These observations may go some way to explain the unsatisfactory nature of the clinical trials of cardiotocography in recent years. They suggest that the scientific basis of visual analysis of FHR variables may be unsound. This does not mean that electronic FHR analysis in labor is useless. Its introduction in Oxford in 1976 was associated with a persistent fall in fetal death rate during labor. But the results call for an objective review of the basis of this practice, and of the numerical criteria which may discriminate between normal and abnormal outcome on delivery.

REFERENCES

1.   Trimbos, J.B. and Keirse, M.J.N.C. (1978). Observer variability in assessment of antepartum cardiotocograms. Br. J. Obstet. Gynaecol. 1978; 85:900-906.

2.   Lotgering, F.K., Wallenberg, H.C. S. and Schouten H.J.A. Interobserver and intraobserver variation in the assissment of antepartum cardiotocograms. Am. J. Obstet. Gynecol. 1982; 144: 701-705.

3.   Flynn, A.M., Kelley, J., and Matthews, K. Predictive value of, and observer variability in several ways of reporting antepartum cardiotocograms. Br. J. Obstet. Gynaecol. 1982; 89: 434-440.

4.   Jongsma, H.W., and Eskes, T.K.A.B. Validity of fetal monitoring during labor, results of multi-center study. In Perinatal Monitoring: 4th Progress Report of the Commission of the European Communities, Medical and Public Health Research (van Geijn H.P., ed), 1985: 25-155.

5.   Nielsen, P.V., Stigsby, B., Nickelson, C., and Nim, J., Intra - and Inter-observer variability in the assessment of intrapartum cardiotocograms. Acta Obstet. Gynecol. Scand. 1987; 66: 421-424.

6.   Donker, D.K., van Geijn, H.P., Derom, R., and Duisterhout, J.S., Processing and results of a pilot study on interventions based upon cardiotocographic recordings, in Dalton, K.J. and Fawdry, R.D.S. The Computer in Obstetrics and Gynecology Oxford, England: IRL Press. 1987; 159-165.

7.   Borgotta, L., Shrout, P.E. and Divon, M.Y. Reliability and reproducibility of non-stress test readings. Am. J. Obstet. Gynecol. 1988; 159: 554-558.

8.   Dawes, G.S., Redman, C.W.G., and Smith, J., Improvements in the registration and analysis of fetal heart rate records at the bedside. Br. J Obstet. Gynaecol. 1985;92:317-325.

9.   Dawes G.S., Houghton C.R.S., Redman C.W.G., and Visser G.H.A. (1982) Pattern of the normal human fetal heart rate. Br. J. Obstet Gynaecol. 89:276-284.

10.   Dawes G.S., Moulden M., and Redman C.W.G. (1990) Limitations of antenatal fetal heart rate monitors. Am. J. Obstet. Gynecol. 162:170-173.

11.   Henson G.L., Dawes G.S., and Redman C.W.G. (1984) Characterization of the reduced heart rate variation in growth-retarded fetuses. Br. J. ObstetGynaecol. 91:751-755.

12.   Henson G., Dawes G.S., and Redman C.W.G. (1983) Antenatal fetal heart rate variability in relation to fetal acid-baselstatus at caesarean section. Br. J. Obstet. Gynaecol. 90:512-516.

13.   Smith J.H., Anand K.J.S., Cotes P.M., Dawes G.S., Harkness R.A., Howlett T.A., Rees L.H., and Redman C.W.G. (1988) Antenatal fetal heart rate variation in relation to the respiratory and metabolic status of the compromised human fetus. Br. J. Obstet. Gynaecol. 95:980-989.

14.   Smith J.H., Dawes G.S., and Redman C.W.G. (1984) Low human fetal heart rate variation in normal pregnancy. Br. J. Obstet. Gynaecol. 94:656-664.

15.   Dawes G.S., Zacutti A., Bovinto F., and Zacutti A., Jr. (ed) Fetal Autonomy
       and Adaptation. John Wiley and Sons: Chichester, U.K. (1990).

16.   Pello L.C., Dawes G.S., Smith J., and Redman C.W.G. (1988) Screening of
       the fetal heart rate in early labour. Br. J. Obstet. Gynaecol. 95:1128-1136.

17.   Ingemarsson I., Arulkumaran J., Ingemarsson E., Tambyraja R.L., and Ratnam
       S.S. (1986) Admission test: a screening test for fetal distress in labour. Obstet.
       Gynecol. 68:800-806.

LASERSPECTROSCOPY:
TECHNOLOGY AND THEORETICAL BACKGROUND

S. Schmidt, U. Wagner, S. Spaniol,
N. Helledie, P. Rolfe, D. Krebs.

Women's Hospital, University of Bonn, Germany
Dept. of Biomedical Engineering and Medical Physics
University of Keele, England

INTRODUCTION

The non-invasive determination of biochemical parameters has become an important aspect of intensive care medicine(2). The newly developed monitors for laser spectroscopy (synonima: nirscopy, reflectometry, reflectance spectrophotometry) provide the possibility of spectroscopic monitoring. The principle was first described by Jobsis who reported on the non-invasive measurement of cerebral and myocardial oxygen sufficiency and circulatory parameters by the means of near-infared spectroscopy(1). The introduction of laser-diodes and the development of mathematical procedures as proposed by Peter Rolfe led to configurations of measuring systems that can be used during research routine(3). Such prototypes has been applied in the field of neonatal intensive care, neurosurgery, and foetal surveillance(4). The configuration of the measuring system aims at the calculation of both the intracellular redox-state and relative changes of blood volume.

Physiological background

Hemoglobin, which is contained in red blood cells, serves as the oxygen carrier in blood, while intracellular cytochromes serve as electron carriers in the respiratory assembly system. Both systems have distinctive structures and properties leading to different absorbance characteristics. During the intramitochondrial oxidation and phophorylation process we find the following reactions of electron transport (see fig. 1). The prosthetic group of both systems, cytochromes and hemoglobin, is the so-called heme. The status of the heme-group is responsible for the capacity of the absorbance of light in distinc wave-length, caused by the reversible changes of the central metal atom during electron transfers (5). The essence of respiration is de-energization of electrons liberated from the citrat-cycle. Through the indirect measurement of the oxygen-dependent step of ATP-synthesis, near-infrared laserspectroscopy gives an indication of the intracellular energy status(3).

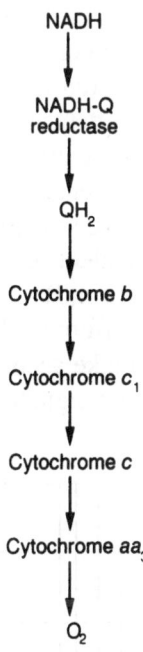

NADH

NADH-Q reductase

QH_2

Cytochrome b

Cytochrome c_1

Cytochrome c

Cytochrome aa_3

O_2

Figure 1: Cytochrome aa3 is the final step of the respiratory chain. Laserspectroscopy detects the redox state by tracing relatively changes of the oxidized status.

Biophysical background

The laser light emitted through a plexi-glass-fibre is not only reflected or absorbed, but in addition it is multi-scattered in the biological sample. The distinctive spectral changes during step-by-step oxygenation of the absorbances are the basis for the measurement by the means of spectroscopy. Differences in the maximum absorbance rate correlates with the concentration of oxidized repectively reduced cytochrome or hemoglobin. While these changes are detectable basically also in the visible range of light, the laserspectroscopy uses radiation in the so-called "optical window" (700 to 1000 nm). In this range the penetration depth of light is high, because the absorbance characteristic of biological tissue represents a minimum in this near-infrared range (see fig. 2). The basis of discrimination between oxidized and reduced cytochrome aa3 is the difference between absorbances at the different laserwavelengths (775-904 nm) according to Rea (see fig. 3)(3). In order to determine the redox state of the cytochrome aa3 in blood perfused tissues, it is not sufficient to measure the absorbance at 845 nm (maximum of oxidized cytochrome aa3). Two parameters have profound effects on the signals, namely the volume and oxygen saturation of the blood in the field of observation (3). During near-infrared laser spectroscopy blood volume can be determined by measuring the attenuation of the transmitted light at 805 nm. Blood oxygenation can be measured at a wavelength corresponding to the absorbance maximum of deoxygenated hemoglobin (775 nm). While the apparent absorbance at 805 nm gives a relative measure of the blood volume, the absorbance at 775 nm is dependent on both the volume of blood and the ratio of deoxygenated hemoglobin(3). The amount of deoxygenated hemoglobin is computed as the difference between the apparent absorbance at 775 nm and that at 805 nm to take account of the effect of blood volume changes on the intensity of 775 nm signal(3).

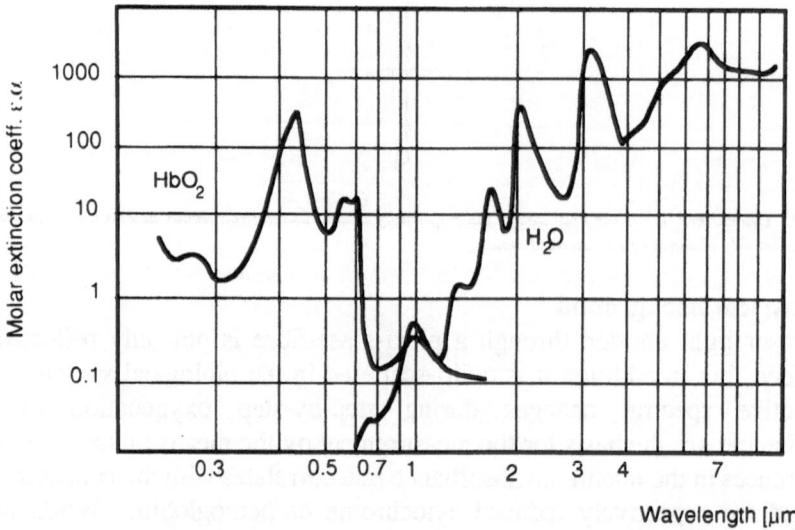

Figure 2: The absorbance of biological tissue is characterized mainly by the absorbance spectrum of the hemoglobine molecule and water. The low absorption range between 700-1000 nm allows a deep penetration of irradiation in this range ("optical window of tissue.")

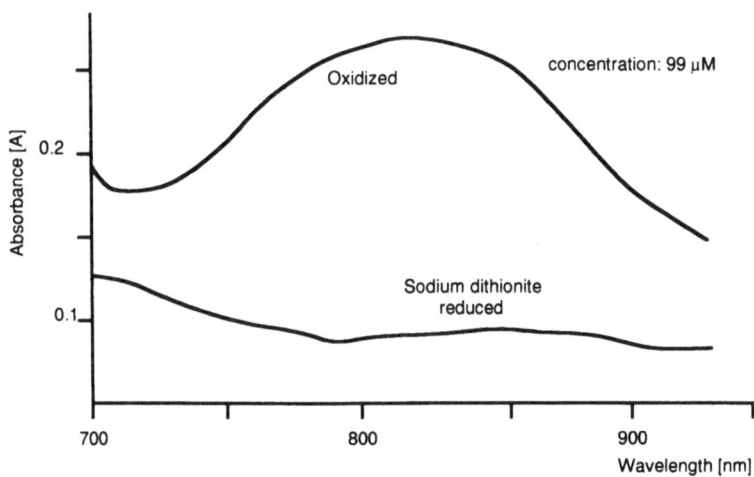

Figure 3: The absorbtion spectrum of cytochrome aa3 is characterized by a peak at 845 nm.

Biomedical engineering-Instrumentation

The Radiometer prototype that was used comprises a personal computer interconnected with a near-infrared data collection unit (NIRDCU). The NIRDCU includes a transmitter board with four laser diodes (wavelengths: 775, 805, 845, 904 nm), as well as a microcomputer board with power supply. Additionally a reciever module with an optical reciever, an amplifier, and a analog-digital converter is installed (Fig. 4).

The light of the laser diodes is conducted by means of optical fiberbundles to a applicator prism located in the sensor. The relected light is collected in a sensor and registered in the NIRDCU by means of photodiodes. Changes of the density in relation to them are displayed on the monitor, while data are stored for processing.

Variations in optical absorbances according to the different wavelengths are the basis for the calculation of relative changes of biochemical parameters being calculated by Rolfe's coefficients. Results of cytochrome aa3, oxigenated hemoglobin, reduced hemoglobin and total blood volume are displayed on the monitor of the connected personal computer (Fig. 4).

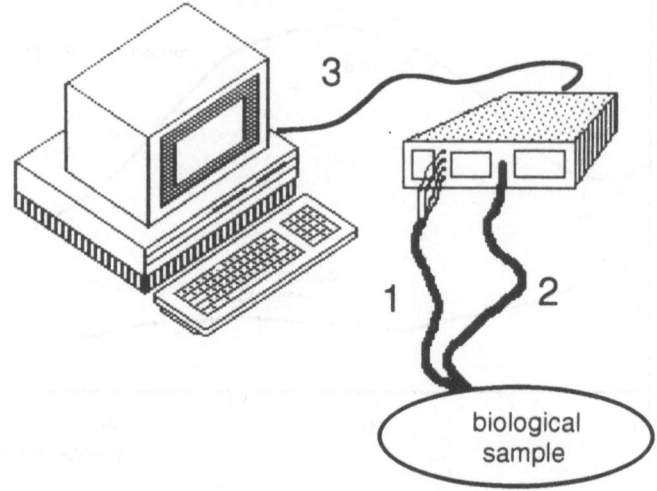

Figure 4: Configuration of the Radiometer prototype for near-infrared laserspectroscopy including a fibre bundle (1) for emitted light from laser diodes, a fibre bundle for conduction of light from the biological sample (2) a data collection unit including an amplifier, and (3) an interconnected personal computer.

REFERENCES:

1. Cady, E.B., Dawson M.J., Hope P.L.: (1983) Non-invasive investigation of cerebral metabolism in newborn infants by phosphorus nuclear magnetic resonance spectroscopy, Lancet 14 May, 1059-1062

2. Jobsis, F.F.: (1979) Oxidative metabolic effects of cerebral hypoxia. Advances in Neurology, 26, 299-318

3. Rea P.A., Crowe J., Wickramasinghe Y., Rolfe P.: (1985) Noninvasive optical methods for the study of cerebral metabolism in the human newborn: a technique for the future? J. Med. Engl. Technol. 9, 160-166

4. Schmidt S., Lenz A., Eilers H., Helledie N., Krebs D.: (1989) Laser spectrophotometry in the fetus. J. Perinatal. Med. 17, 57-62

5. Stryer L.: Oxidative Phosphorylation. in Biochemistry, Freeman New York 1981

CONTINUOUS BASE EXCESS MONITORING IN THE HUMAN FETUS

TOM WEBER AND CARSTEN NICKELSEN

Department of Obstetrics and Gynaecology
University of Copenhagen, Denmark

At present, monitoring of the human fetus during labour is performed either by discontinuous monitoring of the fetal heart rate or by cardiotocography supplemented by discontinuous fetal blood analyses of the acid base state if the technique is available. Although these techniques have been available for more than two decades it has not been possible to improve the low specificity of the methods very much. This have caused many discussions and at present there are a wide scale of recommendations concerning fetal monitoring during labour, a scale running from recommending only auscultation through the recommendation of cardiotocography without fetal blood analysis ending up with some centres recommending routine cardiotocography of all parturients combined with fetal blood analysis in 20-30 per cent of all deliveries.

These problems have emphasized the need for other ways of fetal monitoring. Presently, fetal ECG, analysis of the phono-signal of the fetal heart, and continuous monitoring of the fetal acid base state by electrodes are being investigated as possible improvements in the field of fetal monitoring.

In this chapter we will deal with the possibility of monitoring fetal base excess during labour and the reasons why this parameter is important for the understanding of fetal distress and for the possibility of diagnosing distress before it is too late.

PATHOPHYSIOLOGY OF FETAL ASPHYXIA

In order to understand the mechanism of fetal asphyxia it is necessary to look into three different situations: Fetal metabolic acidosis without hypoxia, fetal respiratory acidosis, and fetal metabolic acidosis caused by hypoxia.

Fetal metabolic acidosis without hypoxia may be caused by maternal transfer of non-volatile acid to the fetus or by medicamentation of the mother. In this situation the pH is low, the pCO_2 normal, the base excess is decreased, and of course the oxygen level is unchanged. In animals, infusion of HCl to fetal lambs results in no brain damage if the oxygen supply is sufficient. Further, if pregnant women are given NH_4Cl just before delivery, a very low umbilical artery blood pH (about 7.05) is not followed by asphyxia (Apgar scores are normal).

Fetal respiratory acidosis may be caused by impaired fetal blood flow to the placenta, impaired placental function, impaired maternal blood flow to the placenta, and increased maternal pCO_2 (e.g. hypoventilation during bearing down efforts). In this situation the pCO_2 increases, pH decreases, and the base excess is normal. In animals, respiratory acidosis with a sufficient oxygen supply does not depress the neonates. However, in humans a very pronounced neonatal respiratory acidosis may induce intraventricular haemorrhage.

Fetal metabolic acidosis caused by hypoxia may be caused by impaired fetal and maternal blood flow to the placenta, impaired placental function, and a decreased maternal oxygen level. In blood you will find a decreased pH, base excess and pO_2, and an increased pCO_2. This is usually the most dangerous type of fetal acidosis as it has been found that the frequency of cerebral haemorrhage is increased in infants delivered in a state of severe metabolic acidosis. Further, monkeys delivered following hypoxic metabolic acidosis develop brain damage very similar to the damage found in human neonates who have died following severe perinatal asphyxia. The brain damage found in monkeys relates to the lactic acid level and pH of blood and cerebral tissue but **not** to the oxygen level of the blood.

A theoretical model of **acid base changes during development of fetal distress** e.g. following compression of the umbilical cord may be divided into two stages:
The first stage is characterized by a decreased oxygen supply to the fetus but all organs are receiving enough oxygen for the metabolism (aerobic metabolism). Consequently, the pH is only slightly decreased because of accumulation of carbon dioxide.
During the second stage of distress the metabolism of some of the fetal organs gradually changes from aerobic to anaerobic metabolism. As carbon hydrate is degraded to only 2 mole of ATP (and 2 mole of lactic acid) during anaerobic metabolism (compared to 38 mole ATP

during aerobic metabolism) the shift towards anaerobic metabolism causes a high consumption of carbon hydrates and accumulation of lactic acid. Further, the accumulation of acids will be buffered by bicarbonate producing carbon dioxide. The result of these changes will be an increasing carbon dioxide tension, and a decreased pH and base excess. In severe cases this accelerating metabolic acidosis may be so severe that all enzyme processes of the fetal cells stop causing cell death.

CALCULATION OF FETAL BASE EXCESS

The base excess of blood is defined as the difference in concentration of strong base between the blood measured and the same blood titrated with strong acid or base to pH = 7.40 at pCO_2 = 40 mm Hg - at a temperature of 37 degrees Centigrade. While the actual base excess is dependent on the buffering capacity of the blood, the standard base excess is an estimation of the base excess of the whole extracellular compartment dependent on the buffering capacity of extracellular fluid.

Three different ways of calculating the base excess are available: The Van Slyke equation, an empirical equation based on computer calculations, and nomograms.

The Van Slyke equation is quite simple - at least compared to other equations. First it calculates the hydrogen carbonate concentration in plasma by the equation:

$HCO_3. = 0.23$ x pCO_2 x $10^{(pH-6.10)}$.
The next step is the calculation of the standard base excess:
$SBE = 0.91$ x $((24-HCO_3.) + 15$ x $(7.40-pH))$.

Empirical equations for calculating the base excess are much more complicated and involves measurement of haemoglobin. They are only used in connection with computer calculations performed during the actual measurements of the blood samples, e.g. by autoanalyzers. If the actual base excess is calculated, the standard base excess can be calculated as the actual base excess at a haemoglobin concentration of 5.8 grams per 100 ml.

Nomograms for reading the base excess have been constructed by Siggaard-Andersen. The principle is that if you know the corresponding values of pH and pCO_2 you can read the base excess, the bicarbonate concentration, and the base deficit on the chart. Further, more complicated acid base charts have been constructed giving the possibility to see what the acid base state is clinically (e.g. acute hypocapnia, chronic base deficit etc.).

MEASUREMENT OF FETAL pH, pCO_2, AND BASE EXCESS IN FETAL BLOOD

Fetal blood sampling has been performed during the last 25 years. If enough blood is sampled in a capillary tube it is possible to analyze all acid base parameters, but often the problem of the method is to sample enough blood without contaminating the blood with air, maternal blood, or amniotic fluid. When the sampling has been performed, looking carefully at the total acid base and oxygenation parameters is very important. This may both give valuable information concerning the state of the fetus and further disclose contamination of the sample (e.g. a high oxygen level and a low level of carbon dioxide is found after air contamination - especially if the haemoglobin concentration is low).

As intermittent sampling of fetal blood is cumbersome to the staff and the parturient - especially if it has to be performed several times - and as the information only reveals the fetal state during one (or a few) short period(s), it is very essential to develop sensors for continuous monitoring of the fetal acid base state.

MEASUREMENT OF THE FETAL ACID BASE STATE BY BIOSENSORS

The only pH electrode which have been thoroughly tested is the Roche glass electrode. It performs very reliable values but it is very difficult to apply, is very expensive, and therefore has not been used during the last 2-3 years. However, it has proved that it is possible to measure fetal tissue pH.

Carbon dioxide may be measured either invasively or non-invasively by transcutaneous electrodes. A transcutaneous electrode (Radiometer) has been tested in a European multicenter study. If 50 per cent of the cases were defined as "completely normal" and the rest as "possibly abnormal", the "completely normal" tracings were always followed by the delivery of neonates without acidosis. However, the 50 per cent being "possibly abnormal" only included very few acidotic neonates. Consequently, transcutaneous carbon dioxide monitoring during labour reduces the need for fetal blood analyses to a certain extent, but the actual pH is not known.

If you combine pH and pCO_2 monitoring during labour, the whole acid base state of the fetus is always known, enabling the obstetrician to reduce the number of obstetrical interventions to a minimum and making it possible to intervene in all cases of real fetal asphyxia.

In a few cases monitoring of the fetal pH by the Roche glass electrode was combined with monitoring the transcutaneous carbon dioxide tension with the Radiometer electrode. Both values were registrated on the cardiotocogram and in seven of the cases on-line calculation of the base excess was performed during delivery. The base excess was either increasing or stable throughout these seven deliveries, all ending with the delivery of infants with Apgar scores of 9-10 and normal acid base state. In one case, simultaneous pH and pCO_2 monitoring was performed while fetal metabolic acidosis developed. It was seen that the pH was decreasing while the carbon dioxide level was constant, thus indicating development of a metabolic acidosis. The infant was depressed at birth but recovered within a few hours.

CONCLUSION

At present, trials with electrodes which were difficult to handle have shown promising results indicating that a combined electrode which can measure both pH and pCO_2 will be a very important step towards managing women in labour safely with a rate of instrumental deliveries very much lower than today.

New pH electrodes currently being tested are either based on the principle of fiberoptic technology or the same principle as the glass electrode but using other materials than glass. These electrodes can be incorporated in a spiral currently being used for cardiotocography. Further, a pCO_2 electrode using the same principle as the pH electrode can be built into the same spiral, the main difference being that the pCO_2 electrode is a pH electrode covered by a carbon dioxide permeable membrane.

When a new combined "acid base electrode" is ready for clinical trial it has to be tested very carefully, hopefully avoiding the problems which arised after introduction of cardiotocography. This means that the electrode should be tested for reliability, and then in randomized multicenter trials comparing it to other methods of fetal monitoring.

IMPORTANCE OF PULSED DOPPLER AND COLOR FLOW MAPPING IN DIABETIC PREGNANCIES

Molly S. Chatterjee, M.D.

University of New Mexico
Department of Obstetrics and Gynecology

Perinatal mortality and morbidity has improved in offspring of diabetic pregnancies. Incidence of congenital malformations has not changed in the last few decades even with improved peripartum management.

Early and accurate prenatal diagnosis of malformations is critical for an informed decision making from ethical point of view and proper perinatal management. First trimester pathology even before appearance of gestational sac can be picked up by application of color flow mapping by demonstration of trophoblastic flow. In diabetic mothers with vascular changes, uterine artery flow velocity wave forms can be studied. Identification of fetal intracranial structures like circle of willis is facilitated by application of color Doppler. Flow velocity wave forms can also be identified with ease and studied from fetal renal, ductal, hepatic and other vital organs. Complex cardiac malformations like atrioventricular canal defects (partial or complete) can be diagnosed more accurately with color Doppler flow mapping. Ventricular Septal Defect (VSD) is the most common form of congenital cardiac malformations in offsprings of diabetic mothers. Small muscular VSD's can only be picked up by color Doppler flow mapping. Early prenatal diagnosis is important for diabetic pregnant patients.

Endovaginal color Doppler has added a new dimension to the detection of fetal malformations. Achievement of euglycemia in the periconceptual period has been a major breakthrough in the management of pregnant diabetics. Endovaginal (EV) color Doppler has new applications in high risk pregnancies. Color Doppler and pulsed wave Doppler provides a unique non-invasive method for evaluating abnormal conditions. Uterine, iliac, and ovarian blood flow patterns can be identified by this modality. We agree with Taylor by saying that the diagnostic benefits of EV color Doppler have moved this technique from the research laboratory to the bed side of the patient.

LITERATURE REVIEW

Doppler monitoring of pregnant diabetics is not new in the literature. Allen, et al, has reported in utero thrombosis and neonatal gangrene in an infant of a diabetic mother (IDM). In utero thrombosis of the brachial artery may be one mechanism which leads to limb reduction defects in IDM. Such vascular problems in utero can best be diagnosed by application of color Doppler. Friedman, et al, studied systolic/diastolic (S/D) ratio in the umbilical artery (UA). About 22 percent had elevated S/D ratio.

Deorari, et al, reported that interventricular septum was significantly thicker and left ventercular mass was significantly greater in IDM. Landon, et al, reported statistically significantly higher S/D ratio of 3.0 or higher in the third trimester of pregnant diabetics of class F to R. In women with vascular disease, elevated S/D ratio was associated with intrauterine growth retardation. Bracero, et al, reported on continuous wave (CW) Doppler velocimetry of umbilical and uterine arteries. Third trimester S/D ratio was significantly higher in poorly controlled diabetics. The number of stillbirths and neonatal morbidity was higher in this group. Uterine artery velocimetry allowed the identification of a patient who developed preeclampsia. Olofsson, et al, noticed that UA pulsatility index (PI) was higher in diabetic patients near term indicating a higher placental vascular resistance. These fetuses developed "fetal distress" in labor. Since fetal distress might be more common in diabetic pregnancy, ultrasonic fetal blood flow measurements are recommended for antenetal fetal surveillance. Bracero, et al, found elevated S/D ratio in diabetic pregnancies with adverse outcome (poor glycemic control, stillbirths and neonatal morbidity).

Walther, et al, attempted to quantitate cardiac output in IDM. Ventricular septal hypertrophy was noticed in 43 percent of IDMs. Morbidity increased with advancing septal thickness. Cardiac output per kilogram diminished linearly with increasing septal thickness secondary to reduced stroke volumes at comparable heart rates. Thus, the information obtained in the literature so far mandates a closer look at the fetus of a diabetic

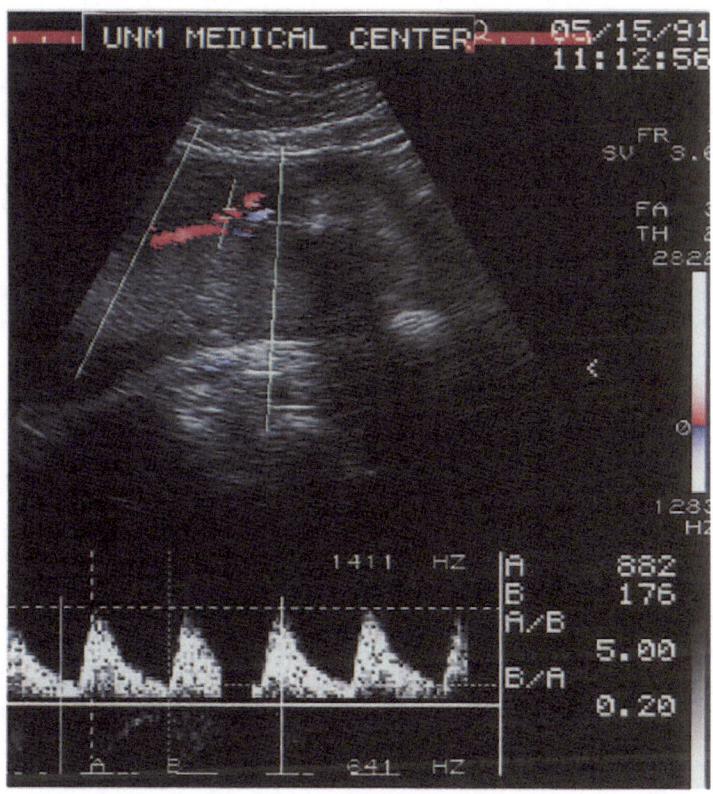

Legend: Color flow mapping and pulse Doppler of fetal of umbilical artery.

mother by more sophisticated color and pulse Doppler studies. Technological advancements in this field should mandate availability of a color Doppler in tertiary care centers.

References:

1)   Van Allen MI, Jackson JC, Knopp RH, Cone R. In utero thrombosis and neonatal gangrene in an infant of diabetic mother. Am J Med Genet 1989; 33:323-7

44

2)   Friedman DM, Ehrlich P, Hoskins IA. Umbilical artery Doppler blood velocity
      waveforms in normal and abnormal gestations. J Ultrasound Med 1989; 8:35-80

3)   Deorari AK, Saxena A, Singh M, Shrivastava S. Echocardiographic assessment
      of infants born to diabetic mothers. Arch Dis Child 1989; 64:21-4

4)   Landon ME, Gabbe SG, Bruner JP, Ludmir J.   Doppler umbilical artery
      velocimetry in pregnancy complicated by insulin-dependent diabetes mellitus.
      Obstet Gynecol 1989; 73:961-5

5)   Bracero LA, Jovanovic L, Rochelson B, Bauman W, Farmakides G.
      Significance of umbilical and uterine artery velocimetry in the well-controlled
      pregnant diabetic. J Reprod Med 1989; 34:273-6

6)   Olofsson P, Lingman G, Marsal K, Sjoberg NO. Fetal blood flow in diabetic
      pregnancy. J Perinat Med 1987; 15:545-53

7)   Bracero L, Schulman H, Fleischer A, Farmakides G, Rochelson B. Umbilical
      artery velocimetry in diabetes and pregnancy.   Obstet Gynecol 1986; 68:654-8

8)   Walther FJ, Siassi B, King J, Wu PY. Cardiac output in infants of insulin-
      dependent diabetic mothers. J Pediatr 1985; 107:109-14

9)   Taylor KJW, Burns PN, Wells PNT. Ultrasound Doppler flow studies of the
      ovarian and uterne arteries. Br J Obstet Gynecol 1985; 92:240-246

A CERTIFIED NURSE-MIDWIFE'S PERSPECTIVE ON INTRAPARTUM BIOCHEMICAL MONITORING OF THE FETUS

Kay Sedler, CNM, MN

University of New Mexico
Department of Obstetrics and Gynecology

Ethical use of resources is essential. Only 40 percent of pregnant women in New Mexico receive adequate prenatal care. This state has ranked worst in the nation in the proportion of women receiving late or no prenatal care. This is a problem in other areas of the United States as well as the world. How many people can we provide prenatal care to for the cost of one machine or one test? We should not do without the technological advances, but we need to keep in mind that these advances are expensive and there must be a balance.

Second, many of you are from University settings and have an impact on teaching medical and midwifery students. It is important for students to learn about technological advances and evaluating and developing research and how to use sophisticated machinery, but we must continue to teach them to use their eyes, ears, and hands also. Many of the physicians, midwives, and nurses we are educating will be in the areas where sophisticated machinery and procedures will not be available. If the providers in the rural areas are able to use these basic clinical skills, they will be able to make appropriate referrals to areas that have access to the technological advances. Again, there must be a balance and I challenge you as teachers and mentors to maintain this balance.

The third issue is risk management. I attended two conferences this year that were about risk management. I was very disappointed at the end of the conferences. The speakers discussed issues relating to standards and documentation and defensive practice. No one discussed including the patient in developing a plan of care and giving them choices and common courtesies. When a women is pregnant, there is a loss of control. Then, what little control a woman has is many times taken away from her by her physician or midwife who develops a plan without her participation.

Women and families want to take some of the responsibility. It is inappropriate to tell a women "We will take care of it for you." We are human and we cannot guarantee perfect babies or perfect labors even if we utilize all our tools including the latest technology. But if you include women in their plans, they share the responsibility and understand there are no guarantees.

Just because you include the woman in the plan of care does not guarantee you will not get sued but it may decrease your chances. At times we get so excited with technology we forget that we are taking care of women and their families. It is a side effect of this excitement that we forget to listen. Dr. John Stone, an Emory University Cardiologist as well as a poet, believes there is a vibrant link between medicine and literature. He believes the patient is a storyteller and the medical condition a text. The thing that unites medicine and literature are the human stories and the chance to hear those stories. Physicians are privileged to hear very compelling ones. We are missing something in medicine and a large part of it is listening to stories.

Another aspect of developing a plan of care is providing choices. There are many ways to do things and you can give the woman a choice. Again, remember that women feel pregnancy is a loss of control. Giving choices, even if they are minor choices, gives the woman a feeling that she is in control of at least one area of her care/life.

Common courtesy -- I hear many complaints about this from patients. Remember, we are not taking care of fetal doppler flow mapping in Room 2 or the ultrasound in Room 4. We are taking care of individual women with names. It takes less than 5 seconds to introduce yourself when you enter a room to do a sophisticated procedure. It takes minimal time to explain what you are doing and to listen to the patient. On 10 percent of CNMs in the US have been named in lawsuits. Part of the reason may be we are willing to share the planning, offer choices, and listen to their stories.

In summary, the important issue is balance -- balancing resources and technological advances, balancing teaching of basic skills and technology, and balancing humanistic care with technology. And, if we listen to the stories it will make it easier to perform this balancing act.

CEREBRAL PALSY AND FETAL HYPOXIA

Luis B. Curet, M.D.

University of New Mexico
Department of Obstetrics and Gynecology

Little, in 1862[1], stressed the association of abnormal parturition, difficult labor, premature labor, and asphyxia with cerebral palsy.

In 1897, Sigmund Freud[2], in his early career as a neurologist, stated that difficult birth in itself in certain cases is merely a symptom of deeper effects that influenced the development of the fetus. One has to consider, he argued, that the anomaly of the birth process, rather than being the causal etiologic factor, may itself be the consequence of the real prenatal etiology. Thus, the controversy about the relationship or lack of between intrapartum events and cerebral palsy was evident before our century.

Over the years, most investigators accepted Little's opinion. Two articles in 1941 were important in strongly suggesting that there was a link between brain damage and perinatal asphyxia. Clifford[3] drew specific attention to the effects of perinatal asphyxia on the brain of infants dying after C.S. performed for abruptio placenta. The particular pathologic findings in those patients were brain swelling and cerebral necrosis. Friedman and Courville[4], in the same year, strongly suggested that perinatal asphyxia led to subsequent development of ulegyria and status marmoratus of the basal ganglia. Subsequent studies by Lilienfield[5], Banker[6], Malamud[7], and others have also emphasized the importance of perinatal asphyxia in the pathogenesis of perinatal brain damage.

Although these general observations were extremely important in establishing the general relationship between perinatal asphyxia and cerebral damage, the relative roles played by asphyxia and other potential factors remained speculative. Nevertheless, it became accepted dogma, particularly in legal circles, that brain damage is a frequent consequence of perinatal asphyxia.

Because of the inherent limitations of studies performed on humans, an understanding of the pathogenesis of perinatal brain damage did not emerge until animal models were developed. Myers[8], in 1972, established a rhesus monkey model that used intrauterine asphyxia to produce damage to the brain of the fetus. Total asphyxia of 10 minutes or less resulted in fetal survival with no recognizable morbidity in the neonate. Total asphyxia of 25 minutes or longer was followed by severe brain damage, however those fetuses died in the immediate neonatal period from myocardial injury resulting from the in-utero asphyxia. In fetuses asphyxiated for 10-25 minutes, damage to brain stem structures was found, however, this type of injury (brain stem) is not seen in the human fetus.

On the other hand, prolonged partial asphyxia resulted in a pattern of brain damage, involving the cerebral hemispheres, similar to what has been observed in humans.

In his experiments, Myers observed:
1.   Late decelerations in association with fetal arterial pH of 7.10 to 7.15. Under these circumstances no damage to the fetal brain was observed in spite of long duration.

2.   That with pH < 7.00 for an extended period of time damage to the fetal brain occurred. These animals, however, died of heart injury.

3.   Lastly, with asphyxia intermediate between 7.00 and 7.10 some animals survived with evidence of brain damage. This represented a small group compared to the other two groups.

Myers' results have been confirmed by other investigators and the general conclusion has been that it is extremely difficult to determine the degree of asphyxia which will cause brain damage and still be associated with survival. Those studies have led to the concept of an all or none effect of asphyxia in the human fetus.

Several clinical studies have confirmed the observation that intrauterine hypoxia/asphyxia can result in brain damage, however, the frequency of the association is still uncertain. In fact, some papers have suggested the infrequency with which the association is seen in clinical practice.

Thus the scientific basis for relating intrapartum asphyxia to subsequent brain damage in an individual patient is tenuous at best.

What events may be responsible for the still unsettled controversy?

1. **From a clinical point of view:**
    Cerebral Palsy appears to be associated with numerous factors other than just asphyxia/hypoxia. Many of such babies suffer from IUGR, thus suggesting pre-existing ante-partal problems. A number of clinical studies have shown that intrapartum hypoxia/asphyxia was followed by neurologic abnormality of the infant only if there was pre-existing chronic fetal distress.

2. **From a developmental point of view:**
    In the developing brain different events will ultimately be expressed with similar tissue pathology. Fetal brain hemorrhage, infection, trauma or asphyxia may all be expressed as cerebral palsy. Even in the absence of definite pathology an alteration in the sequence of events occurring in the developing brain may lead to cerebral palsy.

    In addition, fetal brain injuries in early fetal life may not evoke an identifiable tissue response. Only after 20 weeks can markers of insult and tissue evidence of repair be identified.

3. **From a physiological point of view:**
    Brain damage in an asphyxiated fetus may result from:

    a. Decreased oxygen levels.

    b. Cardiovascular changes induced by the asphyxia and leading to inadequate blood flow through the brain.

Studies in sheep[9] demonstrate that the fraction of the cardiac output (C.O.) directed to the heart and central nervous system equals 0.26 divided by the oxygen concentration ([0]) in the ascending aorta.

When [0] is 4m moles (normal value) 7% of C.O. is directed to those tissues. As the [0] decreases the percentage of C.O. directed to the heart and central nervous system increases. When [0] is 1MM the percentage of C.O. is 26%. [0] below 1 MM create an impossible situation as the percentage required would exceed the amount of blood available.

Additionally, directing blood from the placenta (40% of C.O.) would aggravate the lack of oxygen.

In humans, a greater percentage of the cardiac output is directed to the central nervous system (about six times that of the sheep fetus). Thus, the margin of safety of the human fetal brain is smaller. In other words, the point at which requirements exceed blood flow available would be reached at higher arterial [0]. (See Figure 1)

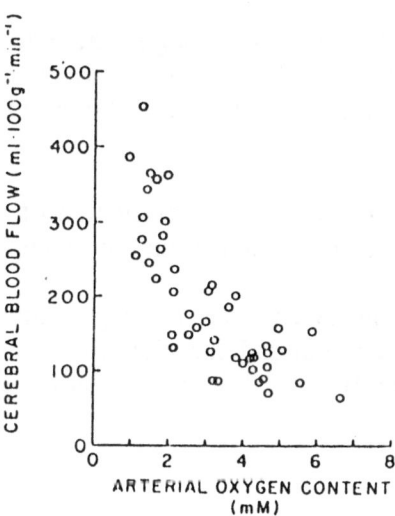

Figure 1: Relationship of fetal cerebral blood flow to arterial oxygen content.

The central nervous system and heart are the only major organs with such an inverse relationship so that the fetus attempts to maintain constant the amount of oxygen carried by the circulation to the brain and heart.

Thus, the fetal circulatory response to hypoxia seems to be based on two requirements:

a.   Increased blood flow to vital organs (CNS and heart).

b.   Maintenance of a constant cardiac output.

Recently, a number of investigators have raised the possibility of brain injury occurring at some point during the antenatal period as a result of a spontaneous umbilical cord complication leading to transient fetal hypoxia/asphyxia not severe enough to cause fetal death but enough to damage critical areas in the developing brain. Volpe[10] has suggested that as a result of cord compression redistribution of cerebral blood flow occurs with hypoxic cellular damage resulting when areas of susceptible white matter are hypoperfused.

These data support the view that oligohydramnios and fetal wasting predisposes to umbilical cord compression which may cause repetitive transient ischemic episodes in the fetal brain leading to neurologic handicap.

Until the advent of ultrasonographic evaluation of the fetus it was not possible to diagnose potential antepartum cord complications.

Use of doppler velocimetry enhances the opportunity to study this issue. Tyrell[11] has shown the use of end diastolic velocity as a predictor of fetal hypoxia. (See Figure 2)

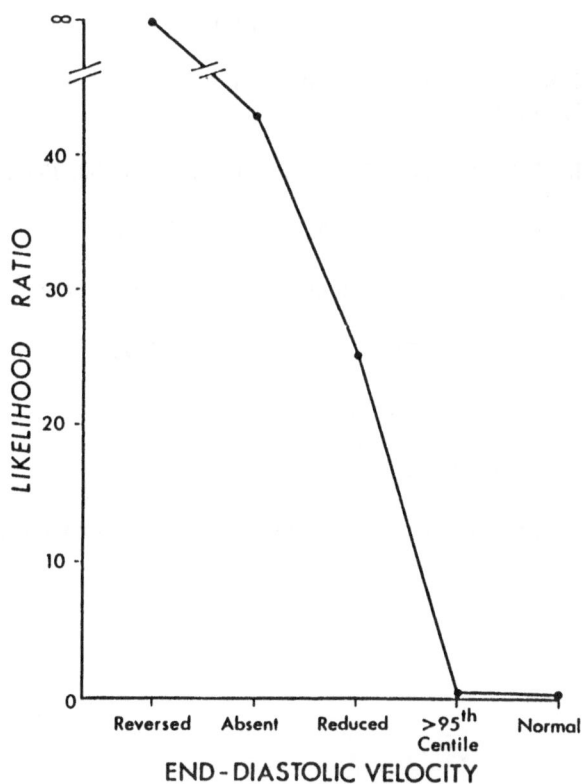

Figure 2: Modified receiver operative characteristics of the Doppler waveform as a predictor of fetal hypoxia.

Thus, further studies utilizing pulsed doppler must be developed in order to clarify this area. Long term prospective controlled studies of motor and mental development of neonates selected because of antenatal abnormal fetal doppler studies must be done in order to ascertain if indeed the so called umbilical cord compression theory is valid.

REFERENCES

1.   Little, W. J. Trans Obstet Soc London 3:293, 1862.

2.   Freud, S. University of Miami Press, 14:1968.

54

3.   Clifford, S. H. J Pediat 18:567, 1941.

4.   Friedman, S. P. and Courville, C. B. Bull Los Angeles Neurol Soc 6:32, 1941.

5.   Lilienfield, A. and Parkhurst E. Am J Hyg 53:262, 1951.

6.   Banker, B. Q. Dev Med Child Neurol 9:544, 1967.

7.   Malamud, N. J Neuropathol Exp Neurol 18:141, 1959.

8.   Myers, R. Am J Obstet Gynecol 112:246, 1972.

9.   Meschia, G. Amer J Obstet Gynecol 132:806, 1978.

10.   Volpe, J. J. Pediatrics 72:589, 1983.

11.   Tyrell, S., Obard, A. H. and Lilford, R. J. Obstet Gynecol 74:332, 1989.

THE EFFECTS OF ANESTHESIA
ON FETAL BLOOD GASES

KAREN KNIERIEM, M.D., Ph.D.

University of New Mexico
Department of Anesthesiology

In order to understand the effects of anesthesia on fetal blood gases, it is important to review the basics from the gross anatomy of uteroplacental circulation to the biochemistry of gas exchange. After reviewing the basics, it will become fairly obvious how anesthesia impinges upon them to potentially alter fetal blood gases.

ANATOMY OF THE UTEROPLACENTAL CIRCULATION

Ten to 20 percent of the maternal cardiac output is directed towards the gravid uterus. The majority of the blood enters the uterus via the uterine arteries and to a lesser extent by the ovarian arteries. These arteries anastomose and penetrate the myometrium as the arcuate arteries. The arcuate arteries bifurcate into the radial arteries which then give rise to the spiral arteries.[1] (Fig. 1) The spiral arteries spurt from the maternal floor of the placenta and shoot like fountains toward the chorionic plate. (Fig. 2) At this point the interface between the two circulations becomes very complex. If the maternal blood intersects the fetal microvilli at an angle, the pattern of flow is called multivillous crosscurrent. If the two circulations are parallel but the flows are in opposite directions, the pattern of flow is countercurrent. When the two circulations are parallel and the flows are in the same direction, it is called concurrent flow. Finally, there are potential areas where maternal blood can pool when in close approximation to the fetal circulation. The geometry of flow becomes very important in regards to gas exchange and will be discussed in more detail later.[2]

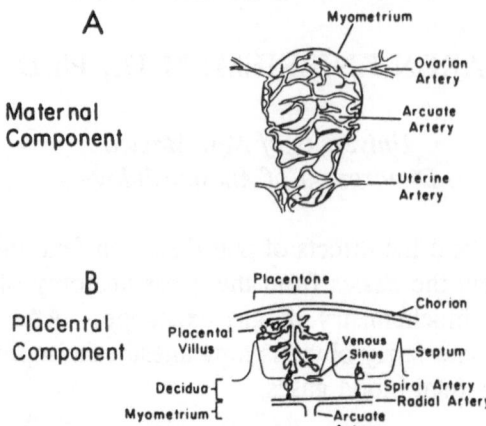

A

Maternal
Component

B

Placental
Component

Figure 1. Blood Supply to the uterus (A) and placenta (B). (Adapted from Burger GA:
Principles of Perinatal Pharmacology. In Ostheimer, GW: Manual of Obstetric
Anesthesia. Churchill Livingstone, Inc., New York, 1984. Used by Permission)

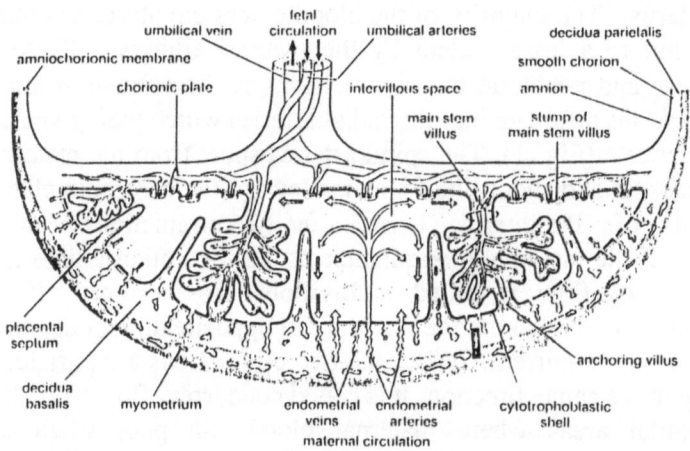

Figure 2. Schematic drawing of a section through a full-term placenta, showing (1) the
relation of the villous chorion (fetal part of the placenta) to the decidua basalis (maternal
part of the placenta), (2) the fetal placenta circulation, and (3) the maternal placental
circulation. Maternal blood flows into the intervillous spaces in funnel-shaped spurts,
and exchange occur with the fetal blood as the maternal blood flows around the villi.
The inflowing arterial blood pushes venous blood out into the endometrial veins, which
are scattered over the entire surface of the decidua basalis. (From Moore KL: The fetal
membranes and the placenta. In The Developing Human: Clinically oriented
embryology. 4th. Ed. WB Saunders, Philadelphia, 1988. Used by Permission)

DIFFUSION AND THE FICK EQUATION

Mechanisms of exchange across the placenta include:

A. Diffusion
B. Active transport
C. Bulk flow
D. Pinocytosis
E. Breaks

Exchange of oxygen and carbon dioxide across the placenta is governed by diffusion which can be defined by the Fick Equation:

$$dQ/dT = KxAxC/D$$

Where:

dQ/dT = rate of transport
 K = permeability
 A = area available for transport
 C = concentration gradient
 D = thickness of the membrane

Permeability (K) is effected by molecular weight, lipid solubility, and electrical charge. If a substance has a molecular weight of less than 1000, the rate is dependent upon the weight. The more lipid soluble and the smaller the electrical charge, the greater the permeability is of any given substance.[3]

The transfer of oxygen across the placenta was initially thought to be permeability limited because the PO_2 of umbilical venous blood (28) was much less than the PO_2 of either uterine arterial blood (96) or uterine venous blood (33). However, subsequent studies have shown that oxygen and carbon dioxide transfer across the placenta is not permeability limited but actually flow limited. The PO_2 of umbilical venous blood is lower than expected for a substance that is flow limited for the following reasons; oxygen is utilized by the placenta, there is ventilation-perfusion (V/Q) mismatching in the placenta, and about 20 percent of the umbilical blood flow is shunted away from the gas exchanging areas of the placenta. This concept of permeability versus flow limited becomes very important when discussing the geometry of the exchanging surfaces and the concentration gradient.[2]

The villous surface area (A) of the human term placenta is about 11 meter^2.[3] However, due to V/Q mismatching and diffusion distances, the area of actual gas exchange is only 1.8 meter^2. In comparison, the adult human lung has a surface area of 70 meter^2,[3].

The component of the Fick equation with the most variability is the concentration gradient. The gradient is dependent upon the concentration of the substance in both circulations, maternal intervillous and fetal placenta blood flow, diffusing capacity, ratio of maternal to fetal blood flow in the exchanging areas, binding of the substance and its dissociation rate, and the metabolism of the substance. The geometry of the exchanging surfaces with regard to blood flow is also important for substances whose diffusion is flow limited such as oxygen and carbon dioxide. The most efficient geometry for exchange is the countercurrent mechanism. This type of geometry would permit the umbilical vein level to approach the level of the uterine artery. The least efficient geometry of exchange is the concurrent mechanism. This type of geometry allows the umbilical vein level to approach the level in the uterine vein. The geometry of flow in the human placenta is considered to be primarily a multivillous crosscurrent system with an efficiency that is halfway between countercurrent and concurrent flows. The umbilical vein PO_2 is lower than expected because of V/Q mismatching, shunts, and placental utilization of oxygen as discussed earlier.[2]

The diffusion distance (D) in the placenta is 3.5 microns. In comparison, the diffusion distance in the adult human lung is only 0.5 microns.[3]

UTERINE BLOOD FLOW AND 0_2 CONTENT

Unless oxygen is transported to the gas exchanging areas, diffusion of oxygen across the placenta cannot occur. Transport is determined by the uterine blood flow (UBF).

UBF = uterine arterial pressure - uterine
    venous pressure / uterine vascular resistance.

Under normal conditions, the uterine vascular bed is almost maximally dilated, and UBF is directly related to the mean uterine arterial pressure. Any decrease in uterine arterial pressure, and/or increase in uterine venous pressure or uterine vascular resistance will decrease UBF and therefore decrease oxygen transport to the placenta. Factors causing a decrease in UBF include; uterine contractions, hypertonus, hypotension, hypertension, and vasoconstrictors, both endogenous and exogenous.

If flow is adequate, oxygen transfer to the fetus is dependent upon relative affinities of hemoglobin, hemoglobin concentration in both circulations, placental utilization, placental surface area, and the distance between the two circulations.[4] In the clinical setting, little can be done to change placental utilization, surface area, or the distance between the two circulations. However, the relative affinities of hemoglobin and the hemoglobin concentrations can be altered either to improve or inhibit oxygen transfer across the placenta.

Fetal hemoglobin has a greater affinity for oxygen than maternal hemoglobin. (Fig. 3) For any given PO_2 fetal hemoglobin will be more highly saturated. When comparing oxyhemoglobin dissociation curves the fetal curve will be to the left of the maternal curve. The PO_2 at which the hemoglobin is 50 percent saturated is 18 for fetal hemoglobin and 28 for maternal hemoglobin. Acidosis can shift either curve to the right, decreasing the affinity for oxygen, and alkalosis can shift either curve to the left, increasing the hemoglobin affinity for oxygen.[5] More important than affinities; however, is the actual content of oxygen in both circulations.

O_2 content = hemoglobin conc. x %sat. x 1.39 + 0.003 X PO_2

Because of a higher affinity for oxygen, fetal hemoglobin will be more highly saturated for any given PO_2. Also, the concentration of fetal hemoglobin (15 gm/dl) is higher than that seen in maternal blood (12 gm/dl).[3] Taking these variables into account, for any given PO_2 fetal O_2 content is much greater than maternal O_2 content. (Fig. 4) It is the difference between these two curves that is the driving force for diffusion across the placental gas exchanging surfaces. Any clinical condition that will decrease the distance between these two curves will adversely influence oxygen transport, and any condition that increases the distance between these curves will improve oxygen transfer. Fetal acidosis and/or fetal hemorrhage will move the fetal O_2 content curve to the right, decreasing the distance between the two curves and adversely effecting oxygen transport. Conversely, maternal alkalosis will shift the maternal O_2 content curve to the left achieving the same adverse effect even in the presence of a normal fetus. Oxygen content is meaningless; however, unless it is delivered to the placenta for gas exchange.

O_2 delivery = UBF x O_2 content.

In the final analysis, both UBF and O_2 content are very important for oxygen transport across the gas exchanging surfaces of the placenta.

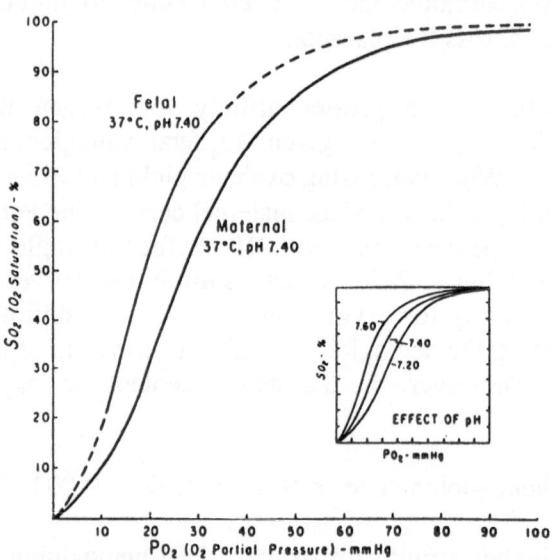

Figure 3. Maternal and fetal hemoglobin-oxygen dissociation curves. The fetal hemoglobin-oxygen dissociation curve is displaced to the left because of the greater affinity for oxygen of fetal hemoglobin. The effect of pH on the position of the curve is shown in the insert. (From Novy, MJ, and Edwards, MJ. Respiratory problems in pregnancy. Am J Obstet Gynecol 99: 1024-1045, 1967. Used by Permission)

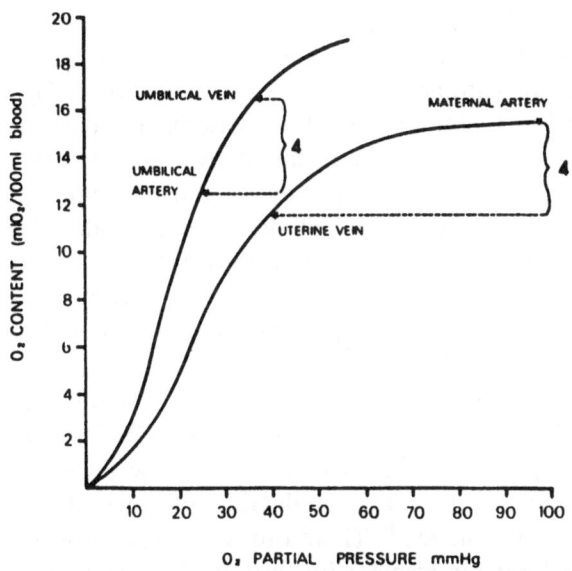

Figure 4. Oxygen contents and tensions, and arteriovenous oxygen concentration differences (brackets), on the fetal and maternal side of the placenta. These are probable values in the undisturbed human. (From Parer, JT. Uteroplacental circulation and respiratory gas exchange. In Anesthesia for Obstetrics, 2nd Ed. S.M. Shnider and G. Levinson, eds. Williams and Wilkens, Baltimore, 1987. Used by Permission)

ANESTHESIA AND FETAL BLOOD GASES

The effects of anesthesia on fetal blood gases are primarily due to the effects of anesthesia on UBF. In 1979, Shnider, et al., applied nonpainful noxious stimuli to awake pregnant ewes and measured plasma norepinephrine levels and UBF before, during, and 10 minutes following the stimuli. They found on the average, a 25 percent increase in plasma norepinephrine levels and a 50 percent decrease in UBF during application of the stimuli. All levels returned to baseline 10 minutes later.[6] Levinson, et al., 1974, hyperventilated awake pregnant ewes through a tracheostomy. During the initial period of hyperventilation, the PCO_2 was decreased to 17. Hyperventilation was continued; however, CO_2 was added to the system to return the PCO_2 to normal levels, then increase it to a PCO_2 of 64. During all periods of hyperventilation UBF was decreased about 25 percent. At no time did the UBF change in relationship to the PCO_2. The conclusions drawn from this aspect of their study were that UBF was not under the influence of PCO_2 as seen in cerebral blood flow, and positive pressure ventilation most likely diminished UBF by decreasing venous return to the right side of the heart thereby decreasing cardiac output. They also looked at PO_2 and O_2 content of maternal and fetal blood during these periods of hyperventilation. They found a significant increase in maternal PO_2 and O_2 content during all periods of hyperventilation. However, during hyperventilation with hypocarbia there was a significant decrease in both fetal PO_2 and O_2 content that returned to control values during hyperventilation when the PCO_2 was normal or hypercarbic. The conclusion drawn from this experiment was that the maternal oxyhemoglobin dissociation curve was shifted to the left decreasing the distance between the maternal and fetal O_2 content curves and therefore decreasing the driving force for oxygen transfer across the placenta.[7] Jouppila, et al., 1977, found a significant decrease of 35 percent in intervillous blood flow in women undergoing elective caesarean section after induction of general anesthesia when compared to baseline levels.[8] Therefore, general anesthesia can adversely effect UBF and oxygen transfer by shifting the maternal oxyhemoglobin dissociation curve to the left through hyperventilation and hypocarbia, releasing catecholamines which decrease UBF, and by decreasing venous return with positive pressure ventilation which decreases cardiac output.

Intervillous blood flow can be increased as much as 35 percent in normal pregnant women undergoing vaginal delivery with a labor epidural.[9] And, in pre-eclamptic patients who have received a labor epidural and whose pressures have been maintained, intervillous blood flow can be increased as much as 77 percent.[10] A labor epidural can also significantly decrease plasma epinephrine levels and potentially improve the labor pattern of highly anxious women who have a dysfunctional labor pattern secondary to the B-sympathomimetic effects of epinephrine.[11] UBF can be decreased by regional anesthesia if the patient becomes hypotensive with the onset of the sympathetic block, or if high concentrations of local anesthesia is injected into the intravascular space.

In 1982, Ramanathan, et al., correlated maternal PO_2 with fetal umbilical artery and umbilical vein PO_2 in a group of women undergoing elective caesarean section under lumbar epidural anesthesia. The patients were divided into four groups (0.21, 0.47, 0.74, 1.00) based on the percent FiO_2 they were breathing. Both umbilical artery and umbilical vein PO_2

correlated closely with maternal PO_2. (Fig. 5) At an FiO_2 of 1.00, the umbilical vein PO_2 significantly increased from a room air value of 15 to 25.

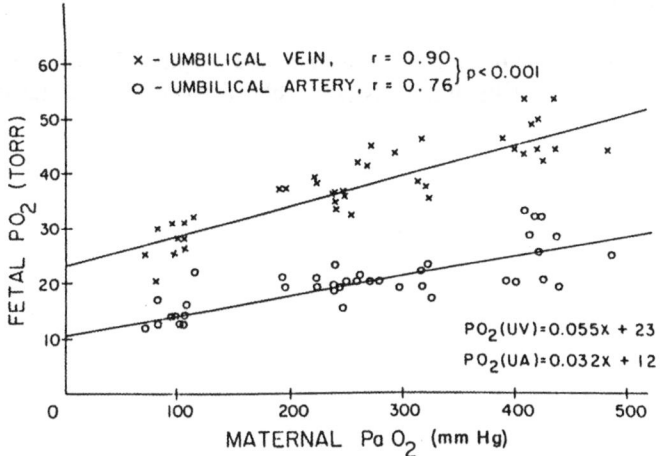

Figure 5. The relationship between maternal and fetal PaO_2. From Ramanathan S., et al.: Oxygen transfer from mother to fetus during cesarean section under epidural anesthesia. Anesth Analg 61: 576-581, 1982. Used by Permission)

In conclusion, optimization of transplacental respiratory gas exchange can be accomplished by avoiding systemic hypotension, increasing maternal PO_2, avoiding maternal hypocarbia, avoiding stress, and avoiding acid-base changes in the fetus.[13]

REFERENCES:

1.   Hunt CO. Uteroplacental circulation - Anesthesiologist's view. In Common Problems in Obstetric Anesthesia, S. Datta and G.W. Ostheimer, eds. Year Book Medical Publisher, Inc., Chicago, 987, pp 58-63.

2.   Metcalfe J, Bartels H, and Moll W.   Gas exchange in the pregnant uterus. Physiol Rev 47: 782-838, 1967.

3.   Parer JT.   Uteroplacental circulation and respiratory gas exchange.   In Anesthesia for Obstetrics, 2nd Ed., SM Schnider and G Levinson, eds.

COLOR DOPPLER AND FETAL ECHOCARDIOGRAPHY

U. GEMBRUCH, R. BALD, M. HANSMANN

Department of Prenatal Diagnosis and Therapy
University of Bonn, Germany

Abstract

Color Doppler flow mapping of fetal heart was performed in 1488 fetuses between 16 and 40 weeks of gestation. Congenital heart diseases were excluded in 1370 fetuses correctly. In 118 fetuses cardiovascular anomalies (without the cases with arrhythmias) were present. In four fetuses, the diagnosis was missed (ventricular septal defect: three cases; coarctation of aorta: one case).

The most important additional informations obtained by color Doppler flow mapping were: (1) diagnosis of insufficiencies of atrioventricular valves; (2) demonstration of turbulent high velocity jet in stenosis of semilunar valve; (3) reverse flow in ascending aorta in aortic atresia and in ductus arteriosus and main pulmonary artery in pulmonary atresia; (4) reverse perfusion of ductus arteiosus and main pulmonary artery along with antegrad high velocity jet in severe pulmonary stenosis as part of tetralogy of Fallot; (5) absent or left-to-right shunt across the foramen ovale in severe obstruction of left ventricular outflow tract combined with mitral insufficiency; (6) bidirectional interventricular shunting of blood in isolated ventricular septal defects and unidirectional interventricular shunting in ventricular septal defects combined with diverse obstructions of ventricular inflow and outflow tract.

INTRODUCTION

Until November 1990, 1488 fetuses were evaluated by color Doppler flow mapping for congenital heart diseases between 16 and 40 weeks of gestation. High risk factors for fetal congenital heart disease were present in 1309 (88,0%) fetuses (Table 1).

TABLE 1: INDICATIONS FOR FETAL ECHOCARDIOGRAPHY

INDICATIONS	No. of CARD. ANOM. = %)
1. Fetal Anomalies	707 (708F) (111=15, 7%)
1.1 Arrhythmia	252 (11=4, 4%)
1.1.1 Supraventricular Extrasystoles	123 (3=2, 4%)
1.1.2 Supraventricular tachycardia	24 (1)
1.1.3 Atrial flutter	3 (1)
1.1.4 Sinusbradycardia	17 (1)
1.1.5 Atrioventricular block	10 (5=50%)
1.1.6 Sinus rhythm (in UFK Bonn)	75
1.2 Sonogr. suspicious cardiac defect	258
1.3 Non-immune hydrops fetalis	109 (25=22, 9%)
1.4 Intrauterine growth retardation	140 (17=12, 1%)
1.6 Extracardiac malformations	148 149F) (25=16, 8%)
2. Risk for cardiac defect	516 (521F) (6=1, 2%)
2.1 Cardiac defect in the family	398 (401) (4=1, 0%)
2.2 Maternal diabetes mellitus	29 (--)
2.3 Teratogenes in 1st trimester	89 (2=2, 2%)
3. Renuncitation of karyotyping (>34 y., low MSAFP)	80 (1=1, 2%)
4. No increased risk	179 (--)

Two-dimensional echocardiography was first performed followed by color Doppler flow mapping. In indicated cases spectral Doppler analysis with pulsed wave Doppler, M-mode-echocardiography and color-coded M-mode-Doppler echocardiography (5) were used. In the most cases the 3.5 MHz sector transducer of an Acuson 128 system was utilized.

For color Doppler flow mapping of fetal heart, the approach is similar to the approach of two-dimensional echocardiography. At first, positions of the fetus and of the fetal heart in the thorax are localized. The viscero-atrial concordance is proofed by visualization of venous inflow in both atria. The interatrial right-to-left shunt across the foramen ovale is demonstrated. Then, the atrio-ventricular connections and the inflow part of the interventricular septum is proofed with the diastolic filling of both ventricles, whose color-filled areas are of similar size. By cranial tilting of the transducer visualization of the left ventricular outflow tract is reached and by further tilting of the right ventricular outflow tract. Thus, the ventriculo-arterial concordance is checked. Finally, aortic arch and ductus arteriosus are demonstrated.

RESULTS

Cardiovascular anomalies (without arrhythmias) were correctly excluded in 1370 fetuses of this high-risk collective. In 118 cases, anomalies were present (congenital heart diseases in 102 cases, and other anomalies in 16 cases (Table 2).

Table 2: CARDIOVASCULAR ANOMALIES (WITHOUT ARRHYTHMIAS) (IN 118 (7.9%) OF 1488 FETUSES)

1. Congenital cardiac malformation	102
1.1 Ebstein's anomaly with pulmonary atresia	5
1.2 Tricuspid atresia	3
1.3 Tricuspid stenosis	1
1.4 Single ventricle	1
1.5 Pulmonary atresia	3
1.6 Pulmonary stenosis and VSD	2
1.7 Tetralogy of Fallot	4
1.8 Atrial septal defect (II)	1
1.9 Ventricular septal defect	14
1.10 Partial AV canal	1
1.11 Complete AV canal	19
1.12 Aortic atresia	3
1.13 Severe aortic stenosis	4
1.14 Hypoplastic left heart	9
1.15 Coarctation of aorta	9
1.16 Interrupted aortic arch	1
1.17 Cor triatriatum and HLV	1
1.18 Double outlet right ventricle	5
1.19 Double outlet left ventricle	1
1.20 d-Transposition of the great arteries	4
1.21 l-Transposition of the great arteries	3
1.22 Truncus arteriosus communis	7
1.23 "Complexe" cardiac malformation	1

Table 2: CARDIOVASCULAR ANOMALIES (WITHOUT ARRHYTHMIAS) (IN 118 (7.9%) OF 1488
FETUSES) (cont...)

2.	Miscellaneous cardiovascular anomalies	16
2.1	Endocardial fibroelastosis (no LVOT-Obstr.)	4
2.2	Aneurysm of left ventricle	1
2.3	Cardiac tumor	2
2.4	Myokarditis	1
2.5	Hypertrophic cardiomyopathy (HCM)	2
2.6	Premature obstruction of foramen ovale	1
2.5	Thoracopagus	2
2.6	Aneurysm of great vein of Galen	3

Exact and complete diagnoses were performed in 105 (87.3%) of these
fetuses. The diagnosis was missed in four cases (ventricular septal defect
in three cases, coarctation of aorta in one case). In six cases, the
diagnosis was incomplete as only the main defect as diagnosed but not all
associated cardiac malformations. In three fetuses, the diagnosis was
incorrect.

The outcome in this series was poor (Table 3) as a result of the high
incidence of fetuses with chromosomal abnormalities, extracardiac hydrops
and intrauterine congestive heart failure.

TABLE 3: "OUTCOME" OF THE FETUSES WITH CARDIOVASCULAR ANOMALIES (N=118)

Anomalies	N	TOP	IUD	NND	INFD	Alive
Cardiac defects	102	36	13	27	4	22 (22%)
norm. karyotype	63	15	8	18	3	19 (30%)
only cardiac defect	38	7	5	11	2	13 (34%)
extracard. malform.	25	8	3	7	1	6 (24%)
abnorm. karyotype	39	21	5	9	1	3 (8%)
Other anomalies	16	5	2	4	-	5 (33%)

In the following part of the results the most important informations of
color Doppler flow mapping are highlighted in addition to two-dimensional
echocardiographic findings.

1. **Insufficiencies of atrioventricular valves**
   Direction, extension of turbulent regurgitation jet can be imaged

degree of valve imcompetence (6). In cases of non-immune hydrops, a pansystolic insufficiency is present, in the other cases, only part-systolic regurgitation. Also, the planimetrically determed jet area seems to be correlated with degree of insufficiency although some theoretical and practical difficulties exist for determing jet area (6,7). Other congenital heart diseases often associated with AV valve incompetences are Ebstein's anomaly, pulmonary atresia type 1, aortic atresia or severe aortic stenosis (8). In these cases, the occurrence of pansystolic mitral insufficiency is correlated with a poor outcome, interatrial left-to-right shunt or complete closure of foramen ovale, and increase of central venous pressure with consecutive occurrence of hydrops (9). In advanced stages of non-immune hydrops fetalis (caused by tachyarrhythmia, anemia, feto-fetal transfusion syndrome, arteriovenous fistula (sacrococcygeal teratoma, arteriovenous aneurysm of great vein of Galen, chorionangioma, and "idiopathic" non-immune hydrops) insufficiencies of AV valves were demonstrated in our series and in other series (5,8).

2. Atresia and stenosis of cardiac valves

In tricuspid atresia filling of hypoplastic right ventricle across ventricular septal defet was seen by color Doppler. In aortic and pulmonary atresia, reverse blood flow from the descending aorta into aortic arch and ascending aorta respectively into ductus arteriosus and pulmonary trunc was demonstrated by color Doppler (9,10). In aortic stenosis and also in cases with pulmonary stenosis an antegrad turbulent high velocity jet was visualized as the sign of stenosis (11). The consecutive spectral Doppler analysis was fascilated by color Doppler. Particularly valuable was the color Doppler in three cases of tetralogy of Fallot with pulmonary stenosis where an antegrad stenotic jet along with a reverse flow via ductus arteriosus into the pulmonary trunc could be demonstrated (10).

3. Intracardiac shunts

Bidirectional interventricular shunting was demonstrated in the cases with isolated ventricular septal defect (4). Small muscular VSD could be demonstrated by color Doppler, even if not detectable in two-dimensional echocardiography. Unidirectional shunting was only present in additional structions of ventricular inflow and outflow.

DISCUSSION

The color Doppler flow mapping is a valuable addition to fetal two-dimensional echocardiography and the prenatal diagnosis of structural cardiac defects and flow abnormalities is facilitated by this method. The accuracy of prenatal diagnosis and, thus prognostic assessment of congenital heart diseases, in particular of complex cardiac defects, are very much increased by the additional information about intracardiac blood flow, thus improving perinatal management. Therefore, in fetal echocardiography of risk-patients, the two-dimensional echocardiography should always be followed by a color Doppler flow mapping.

REFERENCES

1. Allan LD, Crawford DC, Anderson RH, Tynan M: Spectrum of cogenital heart disease detected echocardiographically in prenatal life. Br Heart J 1985;54:523-526.

2. DeVore GR: The prenatal diagnosis of congenital heart disease - a practical approach for the sonographer. J Clin Ultrasound 1985;13-229-245.

3. DeVore GR, Horenstein J, Siassi B, Platt LD: Fetal echocardiography. VII. Doppler color flow mapping: A new technique for the diagnosis of congenital heart disease. Am J Obstet Gynecol 1987;156-1054-1064.

4. Gembruch U, Hansmann M, Redel DA, Bald R: Fetal two-dimensionl Doppler echocardiography (colour flow mapping) and its place in prenatal diagnosis. Prenat Diagn 1989;9:535-547.

5. Gembruch U, Bald R, Hansmann M: Die farbkodierte M-mode-Doppler-chokardiographie bei der Diagnostik fetaler Arrhythmien. Geburtsh Frauenheilkd 1990;50:286-290.

6. Gembruch U, Knoepfle G, Chatterjee M, Bald R, Redel DA, Foedisch HJ, Hansmann M: Prenatal diagnosis of atrioventricular canal malformations using up-to-date echocardiographic technology (a report of 14 cases). Am Heart J 1991;121:1489-1497.

7. Switzer DF, Yoganathan AP, Nanda NC, Woo YR, Ridgway AJ: Calibration of color Doppler flow mapping during extreme hemodynamic conditions in vitro: a foundation for a reliable quantitative grading system for aortic incompetence. Circulation 1987;75:837-846.

8. Silverman NJ, Schmidt KG: Ventricular volume overload in the human fetus: observations from fetal echocardiography. J Am Soc Echo 1990;3:20-29.

9. Gembruch U, Chatterjee M, Bald R, Eldering G, Goecke H, Urban AE, Hansmann M: Prenatal diagnosis of aortic atresia by colour Doppler flow mapping. Prenat Diagn 1990;10:211-217.

10. Gembruch U, Weinraub Z, Bald R, Redel DA, Knoepfle G, Hansmann M: Flow analysis in the pulmonary trunc in fetuses with tetralogy of Fallot by colour Doppler flow mapping; two case reports. Eur J Obstet Gynecol Reprod Biol 1990;35:259-265.

11. Robertson MA, Byrne PJ, Penkoske PA: Perinatal management of critical aortic valve stenosis diagnosed by fetal echocardiography. Br Heart J 1989;61-365-367.

UNM EXPERIENCE WITH COLOR DOPPLER FLOW MAPPING IN FETAL ECHOCARDIOGRAPHY

Raymond R. Fripp, M.D.

University of New Mexico
Department of Pediatric Cardiology

Fetal Echocardiography has become increasingly important in the evaluation of high risk pregnancies and abnormal fetuses. Over a five year period between June, 1985 and May, 1990, 300 fetal echocardiograms were performed in the Pediatric Echocardiographic Laboratory of the University of New Mexico Hospital. These 300 studies were analyzed as to the indications for, and the results of, the studies. Color Doppler was only available in the Laboratory after September, 1989. Selected cases will be shown during this presentation to demonstrate the utility of color Doppler during fetal echocardiography.

Indications for Fetal Echocardiography
225 (75%) of the patients were evaluated to rule out a fetal structural cardiac abnormality. Of these, 94 were performed because of a history of a previous child with congenital heart disease; 105 because of the finding of other structural abnormalities in the fetus (for example, omphalocoele, diaphragmatic hernia, etc.) or functional abnormalities (for example, dilated cardiac chambers) or because of a maternal genetic abnormality (for example, Noonan's syndrome or Marfan's syndrome) or a maternal cardiac abnormality; 26 because of fetal drug exposure secondary to maternal drug ingestion such as Lithium. Suspected fetal arrhythmias were the reason for a fetal echocardiogram in 75 (25%) of the studies.

Technically adequate studies were obtained in greater than 90% of the patients, some however, after more than one evaluation. The length of the studies varied between 30 minutes in easy to evaluate fetuses and 90 minutes in difficult cases. Gestational age of the fetus varied between 16 weeks and 40 weeks.

Results
The fetal echocardiogram was normal in 230 (76%), an isolated

abnormality of cardiac function was detected in 4 (1.3%), an arrhythmia was present in 47 of the 75 patients referred for a possible arrhythmia and a structural abnormality was present in 19 of the 300 studies (6.3%).

The 47 arrhythmias consisted of the following: premature atrial contractions in 30; supraventricular tachycardia in 6; premature ventricular contractions in 3; atrial flutter-chaotic atrial rhythm in 4 and complete heart block in 4.

Structural abnormalities included 5 VSDs, 2 ASDs, 1 aortic stenosis, 3 hypoplastic left heart syndrome, 3 Ebstein's anomaly, 2 hypoplastic right heart syndrome, 2 atrioventricular canals and 1 tetralogy of Fallot.

Of the 225 fetuses screened to rule out a structural abnormality, four of the 94 (4.3%) screened because of a previous child with a cardiac anomaly were abnormal, 17 of the 105 (16.2%) screened because of a fetal structural or functional abnormality were abnormal, one of the 26 fetuses exposed to maternal drugs (3.8%) was abnormal and one of the 75 fetuses with a suspected cardiac arrhythmia was found to have a structural abnormality. Of the 17 fetuses with extracardiac structural abnormalities or cardiac functional abnormalities, 6 had nonimmune hydrops fetalis and of these four (67%) were found to have a structural cardiac anomaly that was probably the etiology of the hydrops.

Discussion
Fetal echocardiography has become an important part of the evaluation of the fetus with extra-cardiac structural abnormalities with 16% of these fetuses demonstrating a structural cardiac abnormality. In the presence of nonimmune hydrops, the frequency of structural cardiac abnormalities is extremely high at 67%. The detection of a cardiac abnormality in association with other abnormalities makes the management of these fetuses more complex and the prognosis poorer. Management of such fetuses is sometimes drastically altered when a significant or lethal cardiac anomaly is detected.

The 4% incidence of cardiac structural abnormalities in fetuses scanned because of a previous child with a structural abnormality is in keeping with the data published regarding the recurrent risk of congenital heart disease in the general population. Despite the low rate of anomalies in this group of pregnancies, fetal echocardiography is worthwhile because of the frequently high level of maternal and paternal anxiety. A family with a previous child who has undergone extensive cardiac surgery has an in depth and graphic understanding of the severity of some congenital cardiac lesions and therefore has an appropriately high level of anxiety and concern regarding the well being of the fetus. The ability to reassure 96% of such families that the fetus is normal must be considered a major benefit of fetal echocardiography. It also allows for counseling and planning of the rest of the pregnancy when a fetal cardiac abnormality is detected.

Sixty-three percent of the studies requested because of concern relating to a fetal arrhythmia confirmed the presence of such an arrhythmia. The majority of these were benign and consisted of premature atrial contractions (30/47) which usually resolved spontaneously before delivery of the infant.

The in-utero diagnosis of a structural abnormality allowed for the planning of delivery of the infant at tertiary care center but in the author's experience was frequently associated with a considerable amount of anxiety and distress on the part of some parents at the discovery that their baby had significant cardiac abnormality.

Conclusion
Fetal echocardiography plays a very useful part in the evaluation of the fetus, especially in the presence of other noncardiac structural abnormalities and may alter significantly the management of the pregnancy and the infant, especially in the presence of nonimmune hydrops. The 4% incidence of fetal cardiac anomalies in pregnancies with a history of a previous of a previous child with cardiac anomalies is in keeping with current data. Cardiac arrhythmias are usually benign and self limiting (premature atrial contractions in 63%) but may occasionally have significant repercussions for the fetus, for example, complete heart block requiring pacemaker implantation after delivery. The advent of color flow

mapping will further enhance our diagnostic capabilities with regard to both structural and functional abnormalities of the fetal heart and will also improve our understanding of fetal cardiac physiology as it allows, in a very graphic fashion, evaluation of fetal cardiac flow patterns in normal and abnormal fetal hearts.

FETAL ARRHYTHMIA:
DIAGNOSIS, SIGNIFICANCE, AND MANAGEMENT

ULRICH GEMBRUCH, RAINER BALD, DIRK A. REDEL, MATTHIAS MANZ, MANFRED HANSMANN

Department of Prenatal Diagnosis and Therapy,
Pediatric Cardiology, and Internal Medicine and Cardiology
University of Bonn, Germany

Fetal arrhythmia is usually first suspected through ausculatory or cardiotocographic findings. The clinical importance of fetal arrhythmia may vary from cases which are benign and self-limited to those that are sustained and associated with fetal congestive heart failure, hydrops and/or death.

The diagnosis of fetal arrhythmia is based on the methods of fetal echocardiography. By two-dimensional echocardiography a structural analysis of the heart can be performed and the most cardiac malformations can be diagnosed prenatally (1,2). Using this approach we are also able to see the signs of congestive heart failure (i.e. skin edema, ascites, pleural and pericardial effusion), that are sometimes associated with long-standing arrhythmia. In some of these polyhydramnios and placental hydrops can also be found. However, an exact diagnosis of the type of fetal arrhythmia is not possible by two-dimensional echocardiography. External fetal heart rate monitoring is inaccurate and virtually useless in the presence of fetal extrasystoles or tachycardia. Only the ventricular rate is registered. Also, the transabdominal fetal echocardiography is of limited value in the analysis of cardiac arrhythmias due to the inability to demonstrate atrial depolarization. Only by M-mode and Doppler echocardiography it is possible to record atrial systole and atrioventricular systolic time relationship which is a prerequisite for accurate diagnosis of fetal arrhythmias (3,4,5).

In M-mode echocardiography with its high time resolution the cursor is localized such that it crosses both atrial wall and ventricular wall or semilunar valves. Thus, the consequences of electrical excitation are of arrhythmia (3,4).

Another possibility is pulsed wave Doppler echocardiography (5). The sample volume is localized in such a way that it registrates ventricular inflow and outflow. Atrial premature beats are demonstrable also by analysis of central venous flow patterns, in particular in the hepatic or inferior caval vein.

Color-coded M-mode Doppler echocardiography is a simulataneous registration of the information of conventional M-mode and Doppler echocardiography (6). Wall or valve movements and also flow phenomena along the cursor are registrated with a high timely resolution. The color-coding is the same as in two-dimensional Doppler echocardiography (7). A flow towards the transducer is blue-colored. The brightness of colors is correlated to the mean Doppler shifts within the sample volumes along the cursor. In variance display a disturbed or turbulent flow is characterized by mixture of green to red or blue and shows a mosaic pattern. The advantage of color-coded M-mode Doppler echocardiography is that the intervals between atrial and ventricular contractions can be analyzed even when the angle of insonation to the fetal heart is unfavorable, since contractions cannot only be identified by wall movements but also by the flow velocities (6).

The most common type of arrhythmia is extrasystolia. About 10% of all fetuses suffer from it, mostly temporarily and with a very good prognosis because of lack of clinical relevance and missing anatomical correlation. The danger is in the initiation of reentry tachycardia which will happen in 1% of the cases (15). Therefore, regular controls of the fetal heart rate should be carried out, e.g. twice a week. In our department 186 fetuses with extrasystoles were diagnosed between 1982 and 1988. 138 of them had SVES, 15 VES, and in 33 cases the extrasystoles have not been classified. During control 17 fetuses showed complications: 11 of them had cardiac malformations, 4 had reentry tachycardia, 1 a QT-syndrome and 1 myocarditis. Out of the population of 186, four fetuses died in utero (three because of cardiac malformation, one because of placental abruption). 79% were normocard before birth. 19% (35 newborns) still showed arrhythmia after birth. Of these, 75% (24 newborns) converted without treatment within the first fourtnight. 11 children had to be treated for varying periods of time.

Five children died postnatally (3 cases of cardiac malformation, 1 case of prolonged QT-syndrome, and 1 case of diaphragmatic hernia).

Another type of arrhythmia is the complete AV block. It is a very rare (incidence about 1:20,000 newborns) but very serious arrhythmia with a bad prognosis if combined with cardiac malformation and/or hydrops (8,9). In our population of 31 fetuses (1981-1989), 17 fetuses developed congestive heart failure in utero already. Nearly all of these had a cardiac malformation, mostly situs inversus with a complete AV channel. Only one fetus with hydrops and a small atrial septal defect survived. In this case complete remission of hydrops was achieved prenatally by transplacental digoxin treatment. All the other survivors never had intrauterine hydrops. If no cardiac malformation was found maternal autoantibodies were usually present. There is still no standard intrauterine treatment for complete heart block. Ventricular rate can be slightly raised by betamimetics (10). In single cases digoxin therapy (9) and ventricular pacing (11) was reported as being successful. With regard to the way of delivery, Caesarian section is not mandatory in non-hydropic fetuses. Intrapartal monitoring can be done by repeated blood samplings, transcutaneous pCO_2 measurement (12) and in the future possibly Laser spectroscopy (13).

The most dangerous arrhythmia is tachyarrhythmia which will mostly be supraventricular tachycardia (SVT), very rarely atrial flutter (AF) (14,15,16). The danger lies in the relatively quick rate which does not allow adequate diastolic ventricular filling (15). This may lead to rapid development of congestive heart failure and hydrops depending on heart rate and duration of the tachycardia. The majority of cases with fetal SVT have reentry tachycardia with a heart rate of 230 to 260 bpm (15). Ectopic pacemakers are very seldom the cause of fetal SVT. Common is a sudden start and ending often triggered by an extrasystole (15). AF with a rate between 360 and 480 bpm is mostly accompanied by 2nd degree AV-block. Hydrops and cardiac malformatons are common (15). In some cases AF will be the result of long lasting SVT, probably due to atrial overdistension. The standard treatment of AF is transplacental digoxin in combination with quinidine (15). First results with direct amiodarone treatment will be reported later. Normalization of the ventricular rate alone is not sufficient. The atrial rate should also be converted (15).

If there are contraindications to immediate delivery, such as fetal immaturity, antiarrhythmic treatment of the fetus should be performed. This therapy is started transplacentally by administration of antiarrhythmic drugs to the mother (14,15,16). For the cases with SVT the drug of first choice is digoxin. Mostly it is necessary to add another drug, as a rule verapamil. When this treatment - high drug dosages are necessary - is ineffective, second line antiarrhythmic drugs should be administered in combination with digoxin. In cases refractory to transplacental treatment additional direct therapy of the fetus may be necessary (17,18).

Since 1981, we have treated 60 cases of tachyarrhythmia in utero. SVT was present in 54 cases, AF in 6 cases. In one case a small muscular VSD was present, in another case multiple rhabdomyomata. 54 of 60 prenatally treated fetuses survived, which equals 88%. All 33 fetuses without hydrops survived. Twenty of 26 fetuses with hydrops survived, which equals 77%.

Since tachyarrhythmias with severe hydrops are often refractory to transplacental therapy, we have early begun to administer drugs directly to the fetus, at first into fetal ascites (17). In the meantime, puncturing of the umbilical vein has become possible. In 1988, we have treated the first case by repeated injections of antiarrhythmic drugs (18). This case showed a generalized hydrops. Transplacental therapy was unsuccessful. Only by direct injections of propafenone into the umbilical vein sinus rhythm could be achieved, but only for periods of 30 to 180 minutes. Meanwhile, the hydrops was increasing. By several samplings, an impaired transplacental passage of digoxin became evident. Fetal to maternal ratios of serum digoxin levels were only 0.2 to 0.3. Therefore, in addition to digoxin amiodarone was administered to the mother. Fetal blood samplings also showed an impaired placental passage of amiodarone. At that time, a fetal loading was performed by repeated injections of amiodarone into the umbilical vein, at first 10 mg, later 40 mg per injection. The periods of sinus rhythm respectively the intervals between injections increased, at first one day, later two days. Altogether, 15 injections of amiodarone were necessary within 22 days. One week after beginning the direct administration of amiodarone, the hydrops decreased and was not detectable at the end of the 30th week. After cardioversion fetal blood samplings showed an increase of fetal to maternal digoxin ratios of 0.7 to 0.8 as expected from the experience within nonhydropic fetuses (18).

In the next case of direct therapy with amiodarone, extensive hydrops, pansystolic imcompetence of both AV valves, complete immobility, and refractoriness to transplacental therapy was present at 27 weeks gestation. After injection of 12.5 mg Amiodarone sinus rhythm could be induced for 10 minutes. At the next control, 15 minutes later, fetal death had to be diagnosed. A placental infarction of 2 cm diameter was seen directly beside the insertion of umbilical cord, at the site of umbilical vein puncture.

In view of this experience, we have reduced the dosages of amiodarone to 2.5 mg/kg estimated fetal weight and increased the dosages to 5mg/kg EW from injection to injection. In the next case following this protocol, we were successful. Eleven injections of amiodarone within 6 days were necessary to achieve a constant sinus rhythm, followed by complete remission of hydrops. In the fourth case SVT was present, but also periods of AF. Panasystolic imcompetence of AV valves could be demonstrated. After 7 injections of amiodarone into the umbilical vein, a constant sinus rhythm was induced, followed again by a complete remission of hydrops. A vaginal delivery could be performed. In the next case at 29 weeks of gestation AF was present. By direct injections of amiodarone long phases of sinus rhythm could be achieved, but no constant cardioversion. At 32 weeks of gestation, Caesarian section had to be performed due to severe preclampsia. This baby had to be treated by thyroid hormones because of hypthyreosis, but only for one month. In the sixth case with SVT refractory to transplacental treatment only two injections of amiodarone were necessary to achieve a constant sinus rhythm.

Our present treatment protocol of fetal tachyarrythmia is summarized in following: If no hydrops is present, transplacental therapy is almost always successful. As a rule, the combination of digoxin and verapamil is enough. Generally, higher digoxin dosages are necessary in pregnancy than in non-pregnant women. When hydrops are present, it is possible that transplacental treatment is ineffective. The transplacental passage of digoxin and possibly of other antiarrhythmic drugs can be severly imparied in cases of hydrops (18). In these cases direct administration of drugs into the fetus is a good alternative. For the umbilical vein injection amiodarone seems to be the drug of choice for three reasons (18): (a) its extremely long terminal elimination half-time reduces the number of umbilical cord punctures required and allows to perform a fetal "loading" even in the absence of tranplacental transfer; (b) amiodarone is highly effective in supraventricular tachycardia and also in atrial flutter (19); (c) amiodarone has a minor negative-inotropic effect compared with other antiarrhythmic drugs (20). However, further experiences in dose selection and drug safety are required, in particular with regard to fetal and neonatal thyroid function (21). As selected dosage of amiodarone 2.5 to 5 mg/kg estimated fetal weight without hydrops seems to be safe and also effective as the last four cases showed.

References

1. Allan LD, Anderson RH, Sullivan ID, Campbell S, Holt DW, Tynan M. Spectrum of congenital heart disease detected echocardiographically in prenatal life. Br Heart J 1985; 523-526.

2. DeVore GR. The prenatal diagnosis of congenital heart disease - a practical approach for the fetal echocardiography. J Clin Ultrasound 1985; 13: 229-245.

3. DeVore GR, Siassi B, Platt LD. Fetal echocardiography. III. The diagnosis of cardiac arrhythmias using real-time directed M-mode ultrasound. Am J Obstet Gynecol 1983; 146: 792-799.

4. Silverman NH, Enderlein MA, Stanger P, Teitel DF, Heymann MA, Golbus MS. Recognition of fetal arrhythmias by echocardiography. J Clin Ultrasound 1985; 13: 255-263.

5. Strasburger JF, Huhta JC, Carpenter RJ, Garson A, McNamara DG. Doppler echocardiography in the diagnosis and management of peristent fetal arrhythmias. J Am Coll Cardiol 1986; 7: 1386-1391.

6. Gembruch U, Bald R, Hansmann M. Die farbkodierte M-mode-Doppler-Echokardiographie bei der Diagnostik fetaler Arrhythmien. Geburtsh Frauenheild 1990; 50: 286-290.

7. Gembruch U, Hansmann M, Redel DA, Bald R: Fetal two-dimensional Doppler echocardiography (color flow mapping) and its place in prenatal diagnosis. Prenat Diagn 1989; 9:535-547.

8. Machado MVL, Tynan MJ, Curry PVL, Allan LD. Fetal complete heart block Br Heart J 1988; 60: 512-515.

9. Gembruch U, Hansmann M, Redel DA, Bald R, Knoepfle G. Fetal complete heart block: Antenatal diagnosis, significance and management. Eur J Obstet Gynecol Reprod Biol 1989; 30:9-22.

10. Kleinman CS, Copel JA, Hobbins JC. Combined echocardiographic and Doppler

assessment of fetal congenital atrioventricular block. Br J Obstet Gynecol 1987; 94: 967-974.

11.   Carpenter RJ, Strasburger JF, Garson A, Smith RT, Deter RL, Engelhardt HT. Fetal ventricular pacing for hydrops secondary to complete atrioventricular block J Am Coll Cardiol 1986; 8: 1434-1436.

12.   Van den Berg P, Gembruch U, Schmidt S, Hansmann M, Krebs D.   Continuous fetal intrapartum monitoring in supraventricular tachycardia by atraumatic measurement of transcutaneous carbon dioxide tension. J Perinat Med 1989; 17: 371-374.

13.   Decleer W, Schmidt S, Gorissen S, Gembruch U, Hansmann M, Krebs D. Kontinuierliche transkutane pCO_2 Messung and Laserspektroskopie zur Geburtsueberwachung bei fetalem AV-Block 3. Grades. Geburtsh Frauenheilk 1990; 50: 227-230.

14.   Bergmans MGM, Jonker GJ, Kock HCLV.   Fetal supraventricular tachycardia. Review of literture. Obstet Gynecol Surv 1985; 40: 61-68.

15.   Kleinman CS, Copel JA, Weinstein TV, Hobbins JC.   In utero treatment of fetal supraventricular tachycardia. Semin Perinatol 1985; 9: 113-129.

16.   Stewart PA, Wladimiroff JW.   Cardiac tachyarrhythmia in the fetus: Diagnosis, treatment, and prognosis. J Fetal Ther 1987; 2: 7-16.

17.   Gembruch U, Hansmann M, Redel DA, Bald R   Intrauterine therapy of fetal tachyarrhythmias: intraperitoneal administration of antiarrhythmic drugs to the fetus in fetal tachyarrhythmias with severe hydrops fetalis Perinat Med 1988; 16: 39-44.

18.   Gembruch U, Manz M, Bald R, Rueddel H, Redel DA, Schlebusch H, Nitsch J, Hansmann M.   Repeated intravascular treatment with amiodarone in fetus with refractory supraventricular tachycardia and hydrops fetalis. Am Heart J 1989; 118: 1335-1338.

19.   Keefe DL, Miura D, Somberg JC.   Supraventricular tachyarrhythmias: Their evaluation and therapy. Am Heart J 1986; 111: 1150-1161.

20.   Schwartz A, Shen E, Morady F. Gillespie D, Scheinman M, Chatterjee D. Hemodynamic effects of intravenous amiodarone in patients with depressed left ventricular function and recurrent ventricular tachycardia.   Am Heart J 1983; 108:848-856.

21.   Foster CL, Love HG.   Amiodarone in pregnancy.   Case report and review of the literature. Int J Cardiol 1988; 20:307-316.

MAGNETIC RESONANCE IMAGING:
A USEFUL ADJUNCT IN EVALUATING
THE ABDOMINAL PREGNANCY

Derek J. Wong, M.D., David A. Turner, M.D.,
James A. Meserow, M.D., Bruce Silver, M.D.,
and Howard T. Strassner, M.D.

Rush Presbyterian-St. Luke's Medical Center
Chicago, Illinois

Magnetic Resonance Imaging (MRI) in obstetrics has been limited to evaluating congenital anomalies, pelvic masses, pelvimetry and placental position. Though ultrasound will continue to be the primary diagnostic tool in evaluating pregnancy, MRI will likely have increased utilization. We report a case of abdominal pregnancy where ultrasound exam diagnosed an intraabdominal pregnancy but had difficulty in accurately localizing the placenta. MRI had an important role preoperatively in evaluating the relationships of the fetus and placenta to the abdominopelvic organs. An MRI was obtained to define the relationships between the intestines, placenta, fetus and pelvic vessels. This study facilitated our surgical approach for delivery of the fetus. Based on this experience, we feel MRI clearly has a role in detailing the anatomic relationships in the surgical approach for an abdominal pregnancy.

Case report

A 42 year old Gravida 3 Para 2 Black female presented at 16 weeks gestational age with lower abdominal pain of one week duration. An ultrasound was performed which demonstrated an intraabdominal pregnancy (Fig. 1). The relationships between the placenta, fetus and pelvic organs were difficult to establish secondary to maternal obesity. After informed consent was obtained, a MRI was obtained which demonstrated the intimate relationship between the placenta and the pelvic anatomy (Fig. 2,3,4,5,). Figure 2 demonstrated a T1 weighted image detailing the placenta, fetus and a placenta free region in the coronal plane. Figure 3 demonstrates a T2 weighted image of the empty uterine cavity. Figure 4 showed a T1 weighted image in the sagittal plane demonstrating the placental free region where the surgical entry was made into the gestational sac. Figure 5 demonstrated the T2 weighted image of the placenta, bladder and endometrial cavity. At laparotomy, a vertical midline incision was performed. Upon entering the peritoneum, the abdominal pregnancy was directly under the mesentary of the sigmoid colon which had firmly adhered to the other pelvic organs. Attempts at removing the mesentary of the sigmoid colon failed. A decision was made to enter the gestational sac by making an opening in the mesentary. The MRI had demonstrated absence of intestines and placenta in this area beneath the mesentary. The gestational sac was entered and the fetus was delivered. Following the delivery of the fetus, profuse bleeding was noted at the edge of the placenta. Abruptio was suspected and attempts to control the bleeding were futile. The abdomen was packed with lap sponges and the patient was brought to radiology for selective embolization of the internal iliac vessels. Following successful arterial embolization of both internal iliac vessels with 700-1000 micron size particles of polyvinyl alcohol and mini coils, the patient was brought to the surgical intensive care. She remained hemodynamically stable. Twenty four hours later, she underwent another laparotomy for removal of the lap sponges but was unable to be weaned off the respirator. A massive pulmonary embolism developed and she subsequently expired on postoperative day 5.

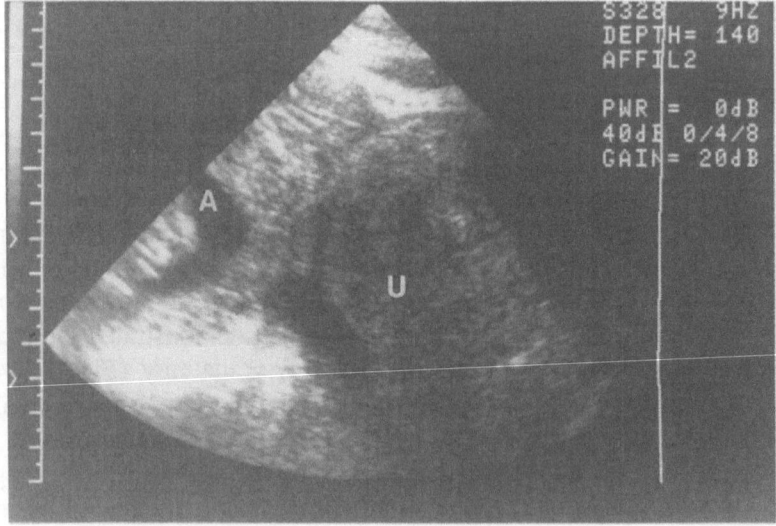

Figure 1: Empty uterus (U) with fundal implantation of intraabdominal pregnancy (A)

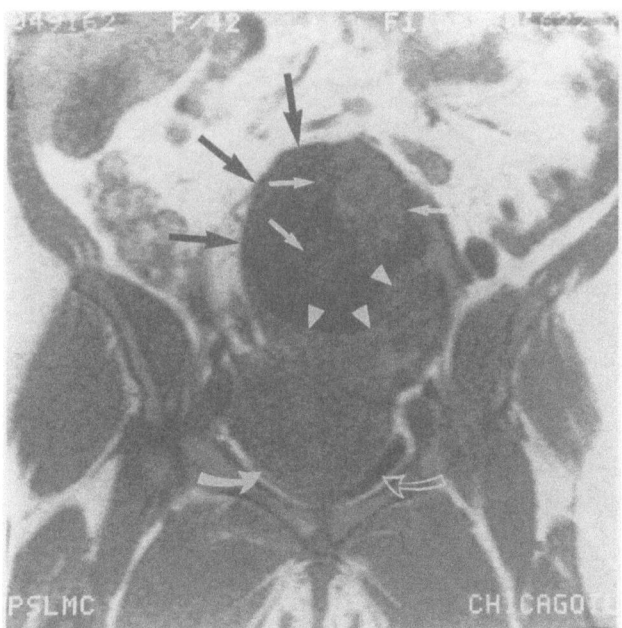

Figure 2: T1 weighted image (coronal section) demonstrating uterus (solid curved white arrows), gestational sac (solid black arrows), fetus (white arrows) and placenta (white arrow head).

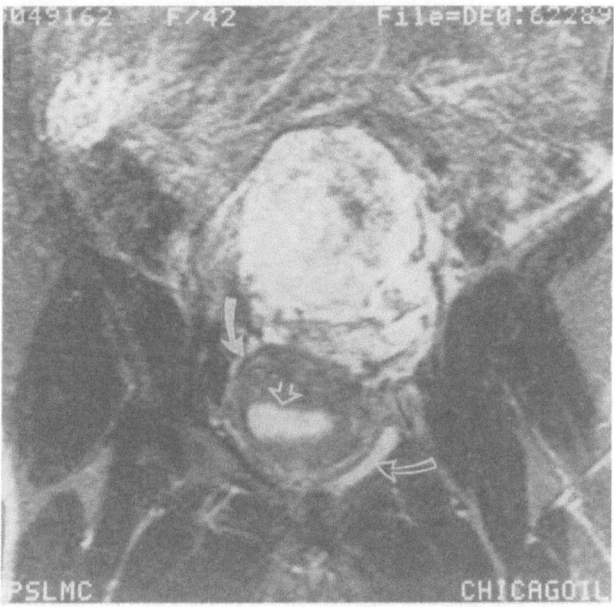

Figure 3: T2 weighted image (coronal section) demonstrating empty endometrium (opened white arrow).

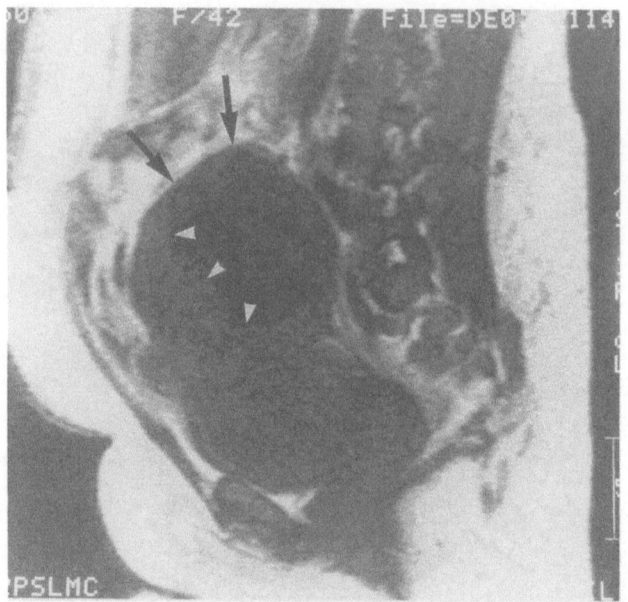

Figure 4: T1 weighted image (sagital section) demonstrating placenta (white arrowheads) and placenta-free region (solid black arrow) where surgical entry was performed.

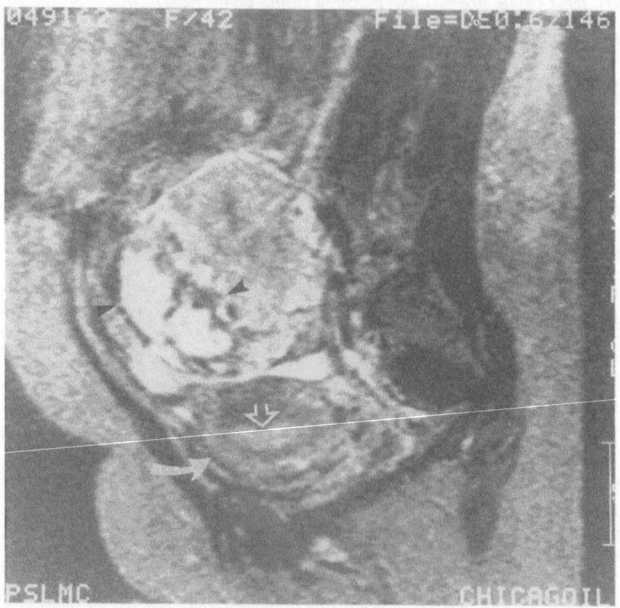

Figure 5: T2 weighted image (sagital section) demonstrating placenta (black arrowheads) empty uterus (open white arrowhead) and bladder (solid white curved arrow).

Comment

The incidence of ectopic pregnancies has been increasing in the United States during the past decade. While the majority of extrauterine pregnancies are intratubal, abdominal pregnancies constitute approximately 1.4% of ectopic pregnancies.[1,2] The abdominal pregnancy is a serious obstetrical complication with a maternal mortality rate ranging from 0.5 - 18% and a perinatal mortality rate between 40-95%.[3] Clinical history and physical examination may be insufficient or misleading in making the preoperative diagnosis. Ultrasound is currently the most effective tool in diagnosing an abdominal pregnancy, however, sonographic interpretations can be difficult due to maternal obesity, overlying bowel gas or distorted pelvic anatomy. Detailed anatomic relationships between the fetus, placenta and maternal pelvic organs and vessels is often difficult or impossible to identify by ultrasound.

Magnetic Resonance Imaging is playing an increasing role in diagnosing and managing complications in pregnancy. Smith reported the first case report of its use in pregnancy in 1983.[4] Since then, numerous reports have detailed the excellent images provided by MRI in evaluating congenital fetal anomalies,[5,6] maternal uterine anomalies,[7,8] placental location,[9] pelvic masses[6] and pelvimetry.[10]

The MRI diagnosis of abdominal pregnancies has been previously reported by Cohen[11], Spanta[12], Hage[13], and Harris[14]. These earlier reports document the clinical usefulness of MRI in diagnosing abdominal pregnancies preoperatively. In this report, the preoperative diagnosis made by ultrasound was confirmed while detailed anatomic relationships were clearly demonstrated in multiple planes with the MRI.

Magnetic Resonance Imaging is unlikely to replace ultrasound as the primary imaging tool in obstetrics and gynecology. Like ultrasound, MRI uses nonionizing radiation and is noninvasive. Unlike ultrasound, however, MRI can provide detailed images in multiple planes without interference from maternal obesity, gas filled or skeletal structures. Ultrasound will continue to be the diagnostic modality of choice for the obstetrician gynecologist due to its low cost and availability. However, MRI is a useful adjunct in evaluating the abdominal pregnancy and can provide additional useful information in planning the surgical approach.

References

1.   Breen JL: A 21 year survey of 654 ectopic pregnancies. Am J Obstet Gynecol. 1970; 106:1004-1019.

2.   Hallat JG, Grove JA. Abdominal Pregnancy: A study of twenty- one consecutive cases. Am J Obstet Gynecol. 1985;152:444-449.

3.   Martin, JN, Sessums JK, Martin RW, Pryor JA, Morrison JC.   Abdominal Pregnancy: Current Concepts of Management. Obstet Gynecol 1988;71:549-557.

4.   Smith FW, Adams A, Philips WP. NMR Imaging in Pregnancy. Lancet 1983;1:61-62.

5.   Williamson, RA, Weiner, CP, Yuh WTC, Abu-Yousef MM. Magnetic Resonance Imaging of Anomalous Fetuses. Obstet Gynecol 1989; 73:952-956.

6.   Weinreb, JC, Lowe TW, Santos-Ramos R, Cunningham, FG, Parkey, R. Magnetic Resonance Imaging in Obstetric Diagnosis Radiology. 1985;154:157-161.

7.   Kelley JL, Edwards RP, Wozney P, Vaccarello, Laifer SA. Magnetic Resonance Imaging to Diagnose a Mullerian Anomaly During Pregnancy. Obstet Gynecol. 1990;75:3(2)521-523.

8.   Yuh WTC, Demarino GB, Ludwig WD, Sato Y, Weiner CP. MR Imaging of pregnancy in Bicornuate Uterus. J Comput Assist Tomog.1988;12:162-165.

9.   Powell MC. Buckley J, Price H, Worthington BS, Symonds EM.Magnetic Resonance Imaging and Placenta Previa. Am J Obstet Gynecol. 1986;154:565-569.

10.   Bryan PJ, Butler HE, LiPuma JP. Magnetic Resonance Imaging of the pelvis. Radiol Clin North Am 1984:22:897-915.

11.   Cohen JM, Weinreb JC, Lowe TW, et al. MR Imaging of a Viable Full-Term Abdominal Pregnancy. Am J Radiol. 1985;145:407-408.

12.   Spanta R, Roffman LE, Grissom TJ, Newland JR, McManus BM.   Abdominal Pregnancy: Magnetic Resonance Identification with Ultrasonographic follow-up of placental involution. Am J Obstet Gynecol. 1987;157:887-889.

89

13. Hage ML, Wall LL, Killam A. Expectant Magagement of Abdominal Pregnancy-A report of 2 cases. Journal of Repro Med. 1988;33:407-410.

14. Harris MB, Angtuaco T, Frazier CN, Mattison DR. Diagnosis of a viable abdominal pregnancy by magnetic resonance imaging. Am J Obstet Gynecol. 1988;159:150-151.

1. Hope JH, WJ LL, Kiffin A. Bayesian Management of Abdominal Pregnancy: a review of 2 cases. Journal of Repro Med. 1988;35:407-410.

2. Balls MP, Anderson TC, Preston GA, Marston DR. Diagnosis of a tubal abdominal pregnancy by magnetic resonance imaging. Am J Obstet Gynecol. 1988;158:156-157.